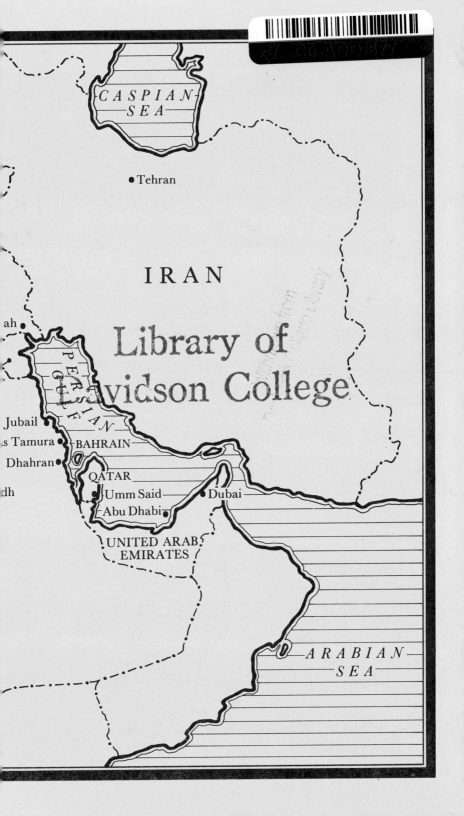

OIL: THE MISSED OPPORTUNITY
or Naft and Shaft

Also by Aubrey Jones

The Pendulum of Politics
(Faber & Faber)

The New Inflation:
the Politics of Prices and Incomes

Aubrey Jones

OIL

THE MISSED OPPORTUNITY
or *Naft and Shaft*

381 PARK AVENUE SOUTH • NEW YORK, N.Y. 10016

First published 1981 by
André Deutsch Limited
105 Great Russell Street London WC1

Copyright © 1981 by Aubrey Jones
All rights reserved

Typeset in Great Britain by
King's English Typesetters Limited, Cambridge
and printed by
Lowe & Brydone Printers Limited, Thetford, Norfolk

ISBN 0 233 97368 0

Contents

	Acknowledgements	7
	Introduction	9
1	Depletion: Fast or Slow?	11
2	The United States: a Conservation Model?	26
3	Canada: the Federation under Strain	38
4	Latin America: Sowing the Oil	50
5	Iran: the Failed Alchemy	67
6	Saudi Arabia: the Conflict within Islam	83
7	Kuwait: the Multi-millionaire City State	101
8	The Smaller States of the Arabian Gulf	118
9	Iraq and Algeria: the Revolutionary Regimes	133
10	The United Kingdom: the Odd Man Out	151
11	The Nature of OPEC	172
12	Depletion Policy and the World Energy Balance	190
13	A Dialogue on Supply	210
	Bibliography	228

Acknowledgements

My first debt is to the United Kingdom Social Science Research Council, whose grant, administered by St Antony's College, Oxford, made possible the maintenance there of a research assistant, Mr Timothy Sweet, and the services of a typist. Mr Sweet is almost solely responsible for the tables in the book and, in particular, for Chapter 12, which expresses a more pessimistic view of the future of oil supplies than those normally put forward; he also helped to keep me on the strait path; if I have strayed the fault is mine, not his. I cannot exaggerate my acknowledgement either to him or to St Antony's College, where I enjoyed many fruitful hours, and the Middle East Centre of which, headed by Mr Robert Mabro, was of enormous help.

Unfortunately, in these days of stringency in Government spending, the grant from the Social Science Research Council did not suffice to cover the foreign travel which I considered essential for the book. For this minimum requirement I relied upon my own resources or the hospitality of most of the countries that I visited, including in the 'countries' the organisations OAPEC (the Organisation of Arab Petroleum Exporting Countries) in Kuwait, and OPEC (the Organisation of Petroleum Exporting Countries) in Vienna. They were uniformly generous and informative, certainly within the limits legally allowed to them. I would like to say a special 'Thank you' to OAPEC, whose objective point of view and constructive attitude proved useful correctives to too quickly gleaned impressions. Their library was extremely helpful, as was also that of the United Kingdom Department of Energy and that of Shell.

Last there are the typists. I do not know whether typing is

a chore or a source of pleasure; I think the answer depends on what is being typed. But I would like to express cordial thanks for their assiduity to Ann Davison and Marlene van Coller.

And, of course, there are my wife and sons, whose apparent forbearance is beyond my understanding.

AUBREY JONES, JUNE 1980
St Antony's College, Oxford

Introduction

Books on oil abound. They deal in the main, however, either with the future relationship between supply and demand, or with the effects on the oil-consuming countries of the steep increases in prices which have taken place since 1973/74. The effects of those same increases on the oil-producing countries themselves is hardly ever considered. It is the purpose of the present book to try to make good that omission.

It attempts to look at the problem of oil from the standpoint of the oil-producing countries, which want to sell oil partly because of the revenues which sale produces, and partly because the world depends on them for oil. They also want to develop – that is, to industrialise. Are they doing one at the expense of the other? It will be the argument in this book that they are. Hence the sub-title 'Naft and Shaft', *naft* being both Arabic and Persian for oil, and *shaft* meaning in Arabic 'suction'. In other words, seen through the eyes of the oil-producing countries, they are being drained of their oil at a rate which threatens their future development. If this contention is true, what is to be the ultimate fate of the oil-producing countries when the oil has run out? Will they indeed have developed, or will they revert to a condition not unlike that prevailing before 1973/74? If the latter outcome is likely, what should be the attitude of the developed world? Can it restrain its appetite for oil? And if it cannot, what is likely to be the effect on price? It will be the argument of this book that the day of oil shortage will arrive earlier than has hitherto been predicted and that the real price will rise.

Finally, if the aim of the oil-producing countries is development, how is this better achieved – through quick depletion or slow depletion, through private initiative or state guidance? It will be the broad conclusion that oil is not easily translated into industrial development, and certainly not without a firm act of government will; otherwise it remains segregated within itself. Thus, a central issue will be the role of government in the oil industry.

In the attempt to answer these questions visits were made to a number of oil-producing countries. Clearly not all could be covered and the choice fell as follows: the United States, as having been the first country to institute a system of control over the rate of depletion; Canada, as a country pursuing a policy of self-sufficiency; Venezuela, as a country wishing to industrialise but having learnt something of the American method of regulating the rate of extraction; Iran, as having, under the regime of the Shah, gone all out for development; Iraq and Algeria as pursuing development while socialistically inclined; Saudi Arabia, Kuwait, Qatar, Bahrain and the United Arab Emirates, as countries more inclined to the adoption of a Western economic system; and finally the United Kingdom, in that, while the problem of Middle Eastern countries is to industrialise, that of the UK is to reindustrialise. What the Middle Eastern countries have done or failed to do may, therefore, carry important lessons for the United Kingdom and, indeed, for other industrialised countries which happen to have oil, but may, like Canada and the United States, be retreating from manufacturing industry.

CHAPTER 1

Depletion: Fast or Slow?

Oil, like human life, is here today, tomorrow gone. It is not, of course, the only product of the earth which can be exhausted. History tells us that the ancient Phœnicians traded in Cornwall for tin. Today there is still some tin left in Cornwall. How much was taken away is simply not known. But when the Phœnicians finally bade goodbye to Cornwall, who cared? A few Celtic tribesmen had been deprived of part of their native wealth, the full value of which they possibly did not know; with the departure of the Phœnicians, that was that.

Centuries later the Spanish conquistadores plundered the golden images found in Mexico and Peru. The images ransacked, the natives were organised into expeditions to seek the sources of the material from which the images had been made, so the alluvial gold of the streams and rivers became exhausted too. Farther south, in what is now known as Latin America, the natives of Brazil did not know of the value of gold, but the Portuguese invaders did. A few centuries later than their Spanish counterparts in the north, they in their turn roamed along the streams and rivers of that vast expanse of territory in search of gold-bearing gravels.[1] Today Brazil can scarcely be reckoned as a major exporter of gold. In 1976 Brazil was the fourteenth largest producer of gold in the world, with just under five per cent of total world output. But who now cares? Who cared earlier? Nobody, for all that was entailed was the robbery of conquered tribes who did not apparently appreciate what they had.

Nearer contemporary times iron ore deposits were found

[1] Gelso Furtado, *Economic Development of Latin America*, Cambridge University Press, 1978, pp. 19, 20.

in hilly districts in the birthland of the Industrial Revolution – the United Kingdom. Iron works, later steel works, grew nearby. The ore was quickly foraged, and when it was gone the steel works which it had brought into being could survive only with ores imported from overseas. It seemed natural that the iron-cum-steel works should be resited at the ports through which ores were now imported, but the inhabitants of the hills beneath which the native ores had first been found were no longer conquered tribesmen. They were an organised working force. They resisted relocation. And the battle to retain the older works – Corby, Ebbw Vale, Consett – has continued to this day, though by and large it has been a losing battle.

The political difference – the difference between conquered tribesmen in distant lands and organised workers in one's own land – causes a different attitude towards those adversely affected. In place of indifference there is at least an element of compassion, masked in the bureaucratic phrase 'redundancy payment'. But has it caused a change in attitude towards the finite materials of the earth themselves? Are we yet convinced of the need to preserve a treasure or at least to exploit it with greater prudence, or perhaps to transform it into another source of wealth, rather than to use it up as rapidly as we can?

The appearance of organised forces of workers has been both preceded, and succeeded, by the creation of nation-states. One nation-state possessing a finite product is Zambia, copper constituting 90 per cent of its export earnings. Copper has benefitted Zambian workers just as iron ore benefitted steel workers in Britain – earnings are high by African standards and some Zambian white collar workers receive earnings comparable to those of their counterparts in Western Europe. However, there is no evidence that the Zambian State has been more conservative in the exploitation of copper than were the Spanish and Portuguese plunderers of South American gold or the iron-masters of Britain in their extraction of native ore. Indeed, the independent State of Zambia may have been no more prudent in its use of

Depletion: Fast or Slow?

copper than the privately owned copper mining companies in the days when Zambia was the British colony of Northern Rhodesia. Two reasons may explain Zambian prodigality: first, a high proportion of the State's tax revenues (50 per cent in 1974) arises from copper; secondly, the high exports of copper are offset by imports of food and agricultural inputs, with the result that there is not enough foreign exchange to import the ingredients necessary to form an industrial society.

State-owned copper in Zambia, then, has been as carelessly raided as any other mineral before it. Nor has it been used to develop the country, though it has helped to create an educational system. Is oil likely to be treated differently? Initially, certainly it was not. That mysterious product, oil – mysterious because there is still some dispute over how exactly it was formed – was first discovered in the United States. According to the law of private property in the United States, many owners may sit on the same field of oil. As will be explained more fully in the next chapter, there is no doubt that competition between owners occupying the same field led to the extravagant extraction of oil – in plain terms to waste. Only gradually did it come to be recognised that there had to be some form of regulation. In addition, when oil was first struck in the United States, that country was already advanced industrially so there was no need to use oil revenues to help along the process of industrialisation. The revenues, though taxed, could be left to those who earned them, no matter how.

The United States has long since ceased to be self-sufficient in oil. From 1950 onwards net imports of crude oil and refined products have increased dramatically. They rose from 0.7 million barrels a day (mbd) in 1950 to 1.7 mbd in 1960 and 8.8 mbd in 1977. Nor is the United States any longer the world's major producer. The balance of production has shifted to the Middle East, to countries destitute of almost everything but oil, countries that in modern parlance are underdeveloped.

Now the nature of underdevelopment may be seen from

two points of view. According to one point of view all nations started from an abject state of poverty, but for a combination of reasons some countries were able to pass through a succession of stages to a high level of development, development being generally synonymous with industrialisation. According, however, to a different point of view, 'evidence is gradually being accumulated that the expansion of Europe, commencing in the fifteenth century, had a profound impact on the societies and economies of the rest of the world. In other words, the history of the underdeveloped countries in the last five centuries is, in large part, the history of the consequences of European expansion. It is our tentative conclusion that the automatic functioning of the international economy which Europe dominated first created underdevelopment and then hindered efforts to escape from it.'[1]

Put differently, 'In the Middle Ages, the Muslim Empires achieved a real flowering of economic life';[2] they became, after the European irruption, exporters of natural resources, some of these resources, like oil, being non-renewable, in exchange for imports of manufactured goods. This was an unequal basis of exchange in that it meant the sale of an asset which could be there in perpetuity for a commodity, either a consumption or a capital good, which was quickly used up. But it was sanctified by the economist's doctrine of comparative advantage, according to which resources were most efficiently distributed when countries exported what they were relatively best at. And the Middle Eastern countries were best at oil. They risked, therefore, seeing their capital asset being drained from them, the drain impeding, if not rendering impossible, development or industrialisation.

Only one new oil-producing country began the process of industrialisation before it became an important exporter of oil – Mexico. 'During the presidency of Lazaro Cardenas, 1939–40, the long dormant revolution of 1910 was revived

---

1 Keith Griffin, *Underdevelopment in Spanish America*, Allen and Unwin, London, 1969, p. 33.
2 Bernard Lewis, *Islam in History*, Alcove, London, 1973, p. 291

by a series of reforms which increased educational opportunities, and changed the political structure of the country. As a climax to these social and economic reforms the government nationalised the railroads and the petroleum industry in 1937–38. Protesting the expropriations, the "imperialist" countries responded with an international boycott. Far from preventing the Mexican "take-off", the boycott accelerated development by forcing the Mexicans to create an efficient entrepreneurial class.'[1]

Development in the sense of industrialisation requires two things: the existence of a 'surplus', and the ability to translate that 'surplus' into industry. The traditional source of the 'surplus' is agriculture. To the French Physiocrats of the eighteenth century the 'surplus' was that part of the agricultural rent received by the landlord which was left over after he had paid his labourers a subsistence wage and provided the seed for next year's crop: the landlord could then use the 'surplus' for his own purposes, either to consume or to invest, possibly in industry.

Apart from the Soviet Union, only one country seems deliberately to have used the agricultural 'surplus' to develop industry – Japan. In 1868 land reforms were introduced which diverted to the Government by way of a land tax the revenues which had previously gone to the landlords. These revenues were used by the State to establish industries and to pay for imported Western technology. After about 1880, when the industries were well established, they were sold off to private owners at relatively low prices so that they could be operated profitably.[2] It was for the same purpose of diverting to industrial investment the 'surplus' of which agriculture was the only source that the Soviet Union collectivised the farms.

The oil-producing countries, particularly those of the Middle East, have had no great agricultural 'surplus'; the land is too arid. But they have had oil, and by its very nature oil provides a substantial economic rent to those who own and

1 Griffin, op. cit., pp. 135, 136.
2 See Tom Kemp, *Historical Patterns of Industrialisation*, Longman, London, 1978.

control it. Given suitable marketing arrangements, oil can be sold for greatly more than it costs to produce, so it is a major source of economic rent. This then has been where the 'surplus' was to be found, and it has grown phenomenally, as is seen from Table 1. The revenues did not initially accrue to the governments. They accrued to the companies to which the governments had conceded territories to explore and exploit. They were like African countries where the establishment of a 'plantation economy brought about a great increase in production, and brought a great flow of money receipts to the colonial region, but the greater part of this was returned to the metropolitan country as profits, as savings of the expatriates and as demand for imports for them.' This '. . . enclave of western business . . . contributed little to the development of the rest of the territory.'[1] Surpluses were obtained from agriculture, mining and raw materials but 'very little of the surplus was devoted to investment in production industry.'[2]

The first few decades of the oil industry in the developing countries saw very little benefit flowing to the local population. There would be some expenditure by the oil companies on infrastructure such as roads, ports and power systems, and these would of course confer an indirect benefit. There would also be some employment of unskilled local labour, mostly for construction. But once the initial construction was completed there would be very little direct link between the oil sector and the rest of the economy. From the viewpoint of an oil company, the stage of development of a country does not affect the company's ability to produce and export oil, provided the infrastructure has been built.

The oil-producing countries found the oil companies uninterested in development, and from the early 1950s onwards the individual governments laid increasing claims to the oil revenues, the process ending almost everywhere in complete nationalisation. The exploration, development and produc-

1 Joan Robinson, *Aspects of Development and Underdevelopment*, Cambridge University Press, 1979, p. 43.
2 Ibid.

Table 1: OPEC member states' annual oil revenues 1950–1978

Annual revenue in millions of current US dollars (a)

	1950	1955	1960	1961	1962	1963	1964	1965	1966	1967	1968	1969	1970	1971	1972	1973	1974	1975	1976	1977	1978(b)
Saudi Arabia (c)	113	288	334	378	410	608(g)	523	663	790	909	927	959	1,214	1,945	2,795	4,340	22,574	25,676	33,500	42,384	36,538
Iran	91	91	285	291	342	380	482	514	608	752	853	923	1,110	1,851	2,396	4,399	17,150	19,030	21,100	21,300	20,500
Iraq	19	207	266	265	266	308	353	368	394	369	569	559	598	986	699	1,843	5,700	7,500	8,500	9,631	9,800
Kuwait (c)	12	307	445	460	480	520	565	600	640	715	700	760	820	995	1,425	1,780	6,545	6,420	8,500	8,900	9,200
Libya					40	108	211	351	523	625	1,002	1,175	1,351	1,674	1,563	2,223	6,000	5,100	7,500	8,850	8,600
Nigeria				19	24	17	20	36	57	60	45	90	247	847	1,117	2,084	5,918	6,570	7,715	9,600	8,200
UAE (d)			4		2	6	12	33	100	110	153	162	212	410	551	900	5,536	6,000	7,000	9,030	8,000
Indonesia (e)												185	245	330	500	685	1,365	3,230	4,500	5,700	5,600
Venezuela	331	596	854	916	1,043	1,078	1,105	1,122	1,099	1,241	1,253	1,256	1,409	1,715	1,912	3,042	9,420	7,095	7,780	6,100	5,600
Algeria			14	28	45	52	60	76	128	199	262	267	272	324	613	988	3,299	3,262	3,699	4,253	5,000
Qatar	1	34	55	54	57	60	64	69	92	102	109	115	122	198	255	400	1,600	1,700	2,020	1,994	2,000
Gabon (f)								1	1	3	4	4	4	9	18	29	173	800	800	600	500
Ecuador															30	129	414	293	533	499	400
Total OPEC												6,455	7,604	11,284	13,874	22,842	85,694	92,676	113,147	128,841	119,938

(a) Includes earnings from exports of refined products and crude petroleum.
(b) Provisional.
(c) Including half of neutral zone, i.e. the zone disputed with Kuwait.
(d) United Arab Emirates, including mainly Abu Dhabi, Dubai and Sharjah.
(e) 1976 to 1978, provisional data.
(f) 1975 to 1978, provisional data.
(g) Includes special payment by Aramco of $153 million.

SOURCES: 1950, Charles Issauri and Mohammed Yeganeh, *The Economics of Middle Eastern Oil*, Praeger, 1962, p. 129; 1955, Petroleum Information Foundation, *Background Information*, No. 8, (1970); 1960 to 1978, Petroleum Economist, *OPEC Oil Report*, Second Edition, 1979, pp. 33 to 39.

tion of oil, with all the expertise and capital layout which that entails, had historically been almost entirely in the hands of the seven major international oil companies (the so-called 'seven sisters'), five of which are based in the United States. In every country the discovery and development of oil had thus been through the medium of foreign companies. It will be seen, in country after country, that the companies failed to contribute to the development of the host community. There was only one way in which the oil countries could achieve development – by claiming a bigger share of the revenues from oil and in controlling the rate of production. This could be done either through very heavy taxation, or through the direct ownership and control of the oil companies – i.e. nationalisation. Events have shown that there was a progression through increasingly heavy taxation to nationalisation, with the oil companies now operating almost everywhere under service or technical assistance contracts.

The relationship between an oil industry and its host nation can be illustrated by reference to a country which will not be discussed in detail in this book: Nigeria.[1] The Nigerian Government provided favourable conditions for the oil companies from the beginning of production in the later 1950s and throughout the 1960s, in an effort to encourage them to discover and develop oil reserves. However, once production was in full swing, the Government tightened the conditions under which the oil companies could operate. This tightening began in 1966, followed in 1969 by a compulsory 51 per cent state participation in all new concessions, and in 1971 by Nigeria's joining the Organisation of Petroleum Exporting Countries (OPEC). The Nigerian National Oil Company was established in July 1971, and from 1972 no new concessions were to be awarded other than to the National Oil Company. Other steps followed to secure a greater share of the oil income for Nigeria, culminating in the nationalisation of British Petroleum in 1979.

Although the Nigerian Government succeeded in securing

1 See S. A. Madujibeya, 'Oil and Nigeria's Economic Development', *African Affairs*, Vol. 75, No. 300, July 1976, pp. 284–316.

a substantial income from oil (oil provided 7 per cent of foreign exchange earnings in 1963 and 86 per cent in 1974), other effects on the economy were limited. Employment in the modern industrial sector is less than 2 per cent of total employment and at present seems unlikely to grow much. There is not much development of industries which use oil or gas as an input, nor is there a significant local manufacture of goods to be used as an input by the oil industry. The oil industry has not had much effect in training manpower, for oil companies find it cheaper to import skills. Part of the justification for a national oil company is that local people will become trained and experienced, and the national company is more likely to spread technology and to become more integrated in the economy. In addition, profits from oil can be retained inside the country.

The changing nature of the relationship between the oil industry and the Government in Nigeria, as the latter tried to secure a greater share of the proceeds from oil, is typical of all oil-producing developing states; and so is the difficulty encountered there in transforming oil wealth into development.

The formation in 1960 of OPEC and in 1968 of OAPEC (Organisation of Arab Petroleum Exporting Countries) were but steps in the process of increasing ownership and control by the host nation. The purpose of OPEC was to determine the price of one kind of oil – light Arabian crude – the prices of other oils being freely determined by reference to this one guiding price. The purpose of OAPEC, on the other hand, was to help promote and, if possible, co-ordinate the development of the Arab countries. Insofar as it was concerned with prices (it has never been concerned with the restriction of production) OPEC was indeed a cartel. Set, however, against the background of underdevelopment, or the failure or inability of the private oil companies to contribute to the development of the host countries which had given them concessions, OPEC was a coalition of nation-states determined to reverse this history.

This aspect of OPEC was ignored by the Western world

when oil prices were quintupled in 1973–74. The reaction of the West was to form the International Energy Agency (IEA), an organisation with a double purpose: first, in the event of a shortage of oil, to share according to need among the participating countries – an innocuous provision; secondly, to act as a countervailing power to OPEC. This latter purpose almost gave the Agency the appearance of a single buyer in opposition to OPEC, the single seller; a fact which put the Agency permanently out of court as a possible vehicle for a future dialogue between oil-producing and oil-consuming countries.

The non-oil-producing countries of the underdeveloped world showed a better understanding of the nature of OPEC than Mr Kissinger, the inventor of the IEA '. . . the poor countries . . . feel a strong pull to the oil-rich countries because of their success in forcing the industrialised world to treat them with deference – an achievement of enormous psychological importance to all developing countries.'[1] In short, OPEC and OAPEC were trying to do what the relevant international organisations, the World Bank, the United Nations Industrial Development Organisation, etc, had failed to do – namely, equalise development by spreading amongst the poorer parts of the world the technologies evolved in the West. They had now got the revenues, but could they use them to industrialise?

This very task faced every individual member of OPEC with an acute dilemma – namely, how much oil to produce now in order to push forward development and how much to leave in the ground for later. These are two faces of the same coin: the faster the rate of development planned, the faster must be the rate of depletion now, assuming no other income; alternatively, the more slowly development is undertaken, the greater the amount of oil which can be left for future generations and possibly the more that can be exported now. Industrialised countries with oil face the same dilemma, but it is more acute for countries not yet industrialised, if only

---

1 Raymond Vernon in *Daedalus,* Fall 1975, p. 249

Depletion: Fast or Slow?

because of the pressure to overtake the West in development. How different countries, industrialised and non-industrialised, have coped with the problem will be described in the chapters that follow.

First, a well-known and widely accepted theory on the subject has to be put to rest. It is a theory with a long pedigree, starting with Hotelling in 1931 and continuing, substantially unchanged, up to the present day.[1] The theory is based on the assumption that the owner of a depletable asset in the ground has another option – investment in another asset, such as stocks and shares, or in another form of production. If the current rate of interest is high, and offers therefore a continuing yield from an alternative investment which promises to be higher than the revenues net of cost expected from the resource in the ground, he will deplete fast, and shift his investment from the asset in the ground to something else. Conversely, if the current rate of interest is low, so that alternative investments are unattractive, and the expected net revenues to be obtained over time from the asset in the ground are likely to be high, he will leave the asset in the ground. In other words, the rate of depletion will depend on the relationship between the current rate of interest and the likely future price of the asset in the ground; a relatively high rate of interest will make for fast depletion, while a relatively low rate of interest will make for delayed depletion.

The theory can best be seen by means of a numerical example. Let us assume that there is a fixed quantity of oil in one field, which, if produced now, would yield $100. If oil

1 H. Hotelling, 'The Economics of Exhaustible Resources', *Journal of Political Economy*, April 1931, No. 39, pp. 137–175.
Other references include: W. J. Baumol, 'On the social rate of discount', *American Economic Review*, September 1968, Vol. 58, pp. 788–802; A. C. Fisher and J. V. Krutilla, 'Resource Conservation, Environmental Preservation, and the Rate of Discount', *Quarterly Journal of Economics*, 1975, Vol. LXXXIX, pp. 358–370; W. D. Nordhaus, 'The Allocation of Energy Resources', *Brookings Papers on Economic Activity*, 1973, No. 3; D. W. Pearce and J. Rose (editors), *The Economics of Natural Resource Depletion*, Macmillan, 1975; R. M. Solow, 'The Economics of Resources or the Resources of Economics', *American Economic Review*, May 1974, Vol. LXIV, No. 2, pp. 1–14. See also *Review of Economic Studies*, December 1974, 'Symposium on the Economics of Exhaustible Resources'.

prices were to rise by 10 per cent per year, ten years hence this oil would yield $100 (1 + 0.1)^{10}$ = $259.4. If the rate of interest were 10 per cent then the oil could be produced this year and yield $100 and that $100, if invested, would similarly yield $100 (1 + 0.1)^{10}$ = $259.4, ten years hence. The producer would therefore be indifferent as between depleting now and depleting later. However, if the rate of interest were only 5 per cent, the total return in year ten would be $100 (1 + 0.05)^{10}$ = $162.9. In this case it is clearly preferable to wait to produce and sell the oil until year ten. Were the rate of interest to be above 10 per cent then the higher return would be achieved by selling the oil now and investing the proceeds in some other form.

The theory sounds plausible enough, but it offers no guide to what happens or what should happen in the real world. In the case of oil the theory overlooks the techniques of good reservoir management. If, in response to a low rate of interest, extraction is low, it can nevertheless be increased later when circumstances change; on the other hand, if there is a high rate of extraction now in response to a high interest rate, the reservoir might well be ruined for ever. It is doubtful whether this eventuality would be in the long-term interest of even a private producer; it would certainly be ruled out of court by a government regulatory authority, and there is no country without some form of government regulation over oil. The reason for this regulation is clear enough; oil is the cheapest form of energy we have or are likely to have for some decades; it has, therefore, no compeer in other forms of productive investment; indeed without it any other form of productive investment would cease to be possible. The assumption, therefore, that there are alternatives to investing in oil is invalid; oil is qualitatively different.

One drawback to the economic theory described above is that the short-term levels of demand, supply and prices produced by the market are most unlikely to be a guide to the best long-term development of the resource. The relative levels of prices and the rate of interest may encourage depletion now, but there can be very little certainty about the values of these variables five years hence, let alone ten years hence, and the time lags in developing new sources of oil and other energy, and in reducing demand through conservation, are so great that the level of prices ten years or more from now are of crucial importance. Thus, it may be to the advantage of oil-producers and of oil-consumers to take decisions about oil which go against the 'signals' of the current market.

The most important weakness of the theory, however, is that oil-producers are now generally governments, anxious to develop their countries, usually in accordance with some plan running at least a few years ahead. The oil revenues, insofar as they are used internally, are invested, for example, in infrastructure (ports, roads, railways) and education. It is very difficult, if not impossible, to compute the return on such investments. But even if it could be done, is it conceivable that the rate of depletion and the resulting investment plans would be changed in accordance with short-term changes in the rate of interest? No, the development plan would go ahead, though it might, of course, be subject to deviation on some other ground.

That this is so is shown by recent events. Between 1975 and 1978, the rate of interest kept roughly in line with the general rate of inflation, while the real price of oil (its price, that is, in relation to the prices of other things) fell. Put differently, the rate of interest rose relatively to the price of oil. The theory would have required some acceleration in the rate of depletion. There is in fact no evidence that there was any such thing.

In reality oil producers do not decide on their levels of production according to the dictates of the theory. The more important determinant of the rate of depletion is the tug between, on the one hand, supplying the world market and thus sustaining the world economy of which the oil-producing countries themselves are part, and, on the other hand, the requirements of domestic development and, to a lesser extent, the development of Third World countries without oil. That the needs of the outside world are taken into account is shown by the fact that when Iranian production fell after the Revolution of 1978–79 from roughly 5 mbd to 3–3.5 mbd, other countries increased their output to make good the shortfall in world supply.

There is no doubt that the outside world does put pressure on the oil-producing countries to produce more oil – the United States presses Mexico and Saudi Arabia, the European Economic Community presses the United Kingdom. Is the outside world claiming so much oil as to impose upon the oil-producing countries too rapid a rate of depletion? And are the oil-producing countries themselves so eager for immediate revenues that they connive in a rapid rate of depletion? Alternatively, have the oil-producing countries been so ignorant of the problems entailed by industrialisation that they have depleted too readily? For industrialisation requires in a traditional society profound changes in education, in social outlook, and indeed in the delineation of national frontiers. Moreover, it has long been a tenet of political science that for economic development to be successful, it must be accompanied by the evolution of a more sophisticated political system and political institutions.[1] If industrialisation is the goal then a tightly controlled rate of depletion is indicated.

The oil-producing countries have been slow to realise this, but it may be the lesson taught by Iran. There is evidence that events in Iran are causing a greater restraint in supply. In that case the industrialised countries will have to put up with the price of oil rising relatively to other prices, a change in their style of living and possibly a decrease in their living standards. As for the oil-producing countries themselves, insofar as they aim at development, the rate of depletion is determined not by forms of investment alternative to oil, but by the length of time required to create the changes which are a pre-condition of industrialisation, and the time-scale is considerable. It may be that even a rate of depletion reduced to match the inevitable slowness of industrialisation will be too

1 See, for example, S. P. Huntingdon, *Political Order in Changing Societies*, Yale University Press, 1968.

high and that the attempt to industrialise will fail. It will be the conclusion of this book that this indeed is likely to be the case, and that at the end of the day oil, like gold before it, will have gone, with nothing to show in its place. That is a generalised conclusion and circumstances will, of course, vary from country to country. Let us then examine how the problem of depletion has been treated in different countries.

CHAPTER 2

The United States: a Conservation Model?

The first country to practise conservation in the ground was the first country to discover oil – the United States. Many features of the policy followed in the United States remain unique to that country, others are of more general interest, while the entire United States treatment of oil reserves has been closely studied by other countries – in particular, by nearby Venezuela.

In Western European countries, what lies beneath the ground has passed, generally as a result of legislation, into the hands of the state. In Islamic countries what lies beneath the ground belongs to the ruler, who, by long custom, concedes possession of the greater part to the state, of which, of course, he is the head. In the United States, however, the sanctity accorded to private property decrees that the owner of a plot of land owns also all that lies beneath it and all that stretches above it, even to infinity – if only he could lay his hands on it.

Imagine then a reservoir of oil underlying two plots of land belonging to different owners. Since both oil and the gas usually associated with it are mobile – that is, they migrate through the porous rock – one of the two owners could suck or drain the oil from that part of the reservoir underlying the land of his neighbour. It would be as though a deer had fled from one of the two plots of land to the other, and on the latter had been slain; in which case the deer would then belong to the owner of the second plot. Justified by this analogy, owners of land over-lying the same field of oil filched from one another, their action being legalised by what was known formally as the 'law of capture'. It was also

The United States: a Conservation Model? 27

informally blessed by the law of the 'invisible hand', according to which a man pursuing his own self-interest, automatically promotes also the interest of the rest of society.

Was this really so? Suppose the owner of a plot of land drained so much oil that he then had to store it in tanks above ground, whereupon the oil evaporated, as often happened, before sale. That would be waste, and waste cannot be in the long-term interest of society. 'The old theory of economics, sometimes called harmonics, according to which spontaneous free market developments are the sole guides to the social weal, has long fallen into disrepute . . . often private profits are made at the expense of appalling long-run social waste.'[1] Because of its geological nature, too rapid recovery from a single oilwell could reduce the total oil ultimately recoverable from that oilfield. Oil is nearly always found in association with saline water and gas. It is the pressure from the expansion of the water and gas which enables the oil, when tapped, to flow to the surface. However, if the oil is withdrawn too rapidly the force of gravity does not maintain an even surface between the gas or the water and the oil, and in consequence pockets of oil can be formed and by-passed, and irrecoverably lost. If oilfields are not carefully exploited as a whole there may be absolute waste of oil. In other words, because competition between different owners of the same oilfield, and even between owners of different fields, makes for as quick a recovery of oil as possible, it also makes for waste.

Oil was discovered in the United States in 1859. The country was then well on the way to industrialisation and did not need the 'surplus' of oil, nor did she need to match the rate of depletion with the requirements of industrialisation. But she did have the problem of waste, and it was not until the approach of the 1930s that there was anything resembling an attempt to deal effectively with it. Who now knows how much oil was dissipated in those intervening eighty years, before the crash of 1929 prompted a change of attitude? Not

1 Erich W. Zimmerman, *Conservation in the Production of Petroleum*, Yale University Press, 1959, p. 126.

that there was an immediate fall in the price of oil. But the massive accumulation of stocks presaged a fall and suggested that it could be averted only through some regulation of production. After decades of doubt the need for regulation came finally to be accepted by producers.

Regulation was undertaken by the individual States, and the method often differed from one State to another. Regulation had two main objectives: first, to protect the rights of owners sitting on the same field and ensure that one owner's reserves were not exploited by another; second, to limit production so as to avoid waste – too rapid exploitation, for example, leading to the premature exhaustion of a field.

The need for the first objective arose from the nature of proprietorial rights in the United States as already described. In an attempt to attain it the regulatory State authority sought to ensure a wider spacing of wells and to secure, as far as possible, the management of each field as a single unit. For example, management could be entrusted to one important owner, provided most of the other owners agreed – 63 or 75 per cent, the proportion varying according to the State. 'Alternatively, or in addition, the unitization agreement may provide for an operating committee on which all owners are represented, each with a vote in proportion to his interest in the unit.'[1]

At first sight it might seem that these arrangements in the United States, arising as they do from the unique nature of proprietorial rights in that country, have little significance elsewhere. This is not so. In the area, for example, of the Persian or Arabian Gulf, 'there are hardly two adjacent countries . . . without the geographic boundary splitting one or more reservoirs.'[2] One such example is the Wafra field in Kuwait, which is shared by Saudi Arabia, and on the whole is badly managed. In North Africa, similarly, there are important fields lying beneath the ill-defined border areas between

---

1 Stephen L. Macdonald, *Petroleum Conservation in the United States*, Johns Hopkins University Press, Baltimore, 1971, p. 200.
2 OAPEC, *Reservoir Engineering*, Kuwait, 1979, p. 24.

Algeria and Tunisia.[1] It is true that under OPEC rules any dispute over claims to the same field is supposed to be referred to the Inter-OPEC High Court. It is doubtful, however, whether any High Court has yet over-ridden the wishes of a sovereign state. In OPEC, as in the European Economic Community, the individual state remains sovereign. In OPEC, therefore, as earlier in the United States, some fields are subject to a beggar-my-neighbour policy, though there are exceptions. One such is a joint field shared by Bahrain and Saudi Arabia, but operated by a single manager – Aramco.

The identical problem has occurred in the North Sea, which, for the purpose of extracting oil, has been allocated or auctioned by the British Government in blocks, these generally being vastly larger than the usual onshore US lease of 160 acres or even the offshore US blocks of about 5,000 acres. Even so, it is possible for several blocks to cover the same oilfield. Accordingly, the British Government has inserted in licences granted a clause requiring adjoining operators to propose a scheme to develop the field as a unit. In default of a suitable scheme put forward by the operators the Government itself has a right to propose a scheme, and any operator dissatisfied with it can go to arbitration. It is not known exactly how these fall-back provisions have operated.

The second objective of control over oil production in the United States has been to even out the rate of extraction. Rapid depletion may in the short run extract more than could be found over a longer spell of time, as explained earlier. For sheer physical reasons, therefore, there is a case for controlling the rate of output. Admittedly the methods of control have been arbitrary. In most cases they have been related to the depth of the well and its distance from the adjoining well. For instance, a producer with a well 4,000 feet deep and drilling over an area of forty acres would be allotted a certain level of production – an arbitrary limitation, clearly. In some few States an attempt has been made to adopt a more refined criterion to govern the rate of output,

1 Muhamad A. Mughraby, *Permanent Sovereignty over Oil Resources*, Middle East Research and Publishing Centre, Beirut, 1966, p. 128.

this criterion being known as the 'maximum efficient rate' (MER), or the rate which over time produces the maximum output. This rate is, however, difficult to define: 'The determination . . . is subject to considerable uncertainty until the reservoir becomes substantially developed and has been in production for some time. Even then some uncertainty remains . . .'[1]

The model set by the United States for controlling the rate of production on physical grounds alone has been followed by many members of OPEC, the control often being more rigorous and being applied to companies publicly owned, whether they encompass part or the whole of the relevant country's oil production. Algeria, Libya, Iraq, Kuwait, and Egypt can all be cited as having instituted, within the Ministry of Energy, a strict controlling unit over the production rates of nationalised companies.

If there is a case for controlling the production of an oilfield on 'physical' grounds, e.g. to prolong the field's life and increase its ultimate yield, there is also a case on economic grounds. Suppose the price is at the moment high, and in the absence of regulation, producers respond enthusiastically, stocking above the ground more oil than can be transported and exposing it to evaporation. This, too is 'waste'. Accordingly most of the North American States – the exceptions are Mississippi, Illinois and Wyoming – attempt to equate production with demand as foreseen. A forecast of the demand for the oil of a particular State for a month ahead is made by the Federal Bureau of Mines, though a State is free, if it wishes to, to differ from the forecast; indeed a State often takes into account other considerations.

The objective of this 'prorationing' of supplies (or allocation of production quotas among producers) has been to prevent prices from falling when supplies outstripped demand. The limitation of production to market demand was in full operation by the 1930s, when, it must be

---

1 *Reservoir Engineering*, op. cit., p. 75

remembered, the United States was self-sufficient in oil. There were thus no imports to undercut domestic prices. The simple objective of 'prorationing' was price fixing, and thus in practice it meant the sale of oil at that price which could be achieved by restricting production to market demand. In effect this was a state-run cartel, the price being fixed by varying the output. However, it also had the largely unintended effect of conserving supplies of oil, and thus could be regarded as a form of depletion policy.

One of the most important bodies concerned with 'prorationing' was the Texas Railroad Commission. This institution has long had considerable powers over the production and pricing of oil in Texas, which, as the major oil-producing State, has thus had great influence on the industry throughout the USA. From the 1930s to the 1960s, the Commission restricted the number of days of permitted production from wells drilled in Texas. The intention and the effect was to prevent supplies of oil from exceeding demand, and thus to maintain the price. Subsequently, OPEC countries have quoted the Texas Railroad Commission as an example of and justification for raising prices.

There has been much argument over the relative benefits and disadvantages of 'prorationing'. On the one hand, there have been ranged those who contend that the limitation of output to foreseen demand makes the short-term price to the consumer higher than it otherwise would be (which indeed it does); that it raises costs, in that wells are made to produce at below their technical optimum; and that it prompts the sinking of wells which otherwise would not have been sunk and which can, therefore, operate only at higher cost. On the other hand there are those who point out that if output were allowed to exceed demand, those producers with inadequate pipelines would have to store above ground, thus entailing great physical waste; that wells nearing the end of their lives (known as 'stripper wells') would have to be prematurely shut down, because of the falling price, with a consequent great loss in the total oil potential of the country; and that conservation and the long-term interest of the consumer would be damaged. Indeed 'stripper wells' have been exempt from 'prorationing'.[1]

Thus, 'prorationing' as practised in the United States has operated in a manner quite different from that of OPEC. Its primary aim has been the control of production, a control which has had the indirect effect of maintaining the price. OPEC, by contrast, has sought to keep up the price, but without any control over production.

The attempt to regulate production according to demand has naturally, in the classic land of free enterprise, given rise to controversy; but it has prevailed, and it has spread its ripples overseas. After the First World War, which gave rise to a fear of the exhaustion of supplies in America itself, American companies were nudged into going abroad. They went to the Middle East, flowing with cheap oil. Yet 'the rate of production [there] . . . was, until the later 1950s, surprisingly closely adjusted to the rate at which it [the crude oil] could be processed and sold as products without "disorderly" effects on prices.'[2] In a sense it could be said that the regulation of production by the companies prevented waste, but the US Federal Trade Commission, in its report 'The International Petroleum Cartel', failed to show that this adaptation to demand was the result of central co-ordination. It was the result, rather, of the fact that the concessions were large and were jointly operated by large vertically integrated companies which extracted just enough crude oil to meet their downstream or end requirements. It was to keep prices stable that the rate of extraction was regulated; and the stability of prices in turn facilitated conservation. Towards the end of the 1950s the independents – that is, producers of oil without further processing facilities – stepped into the market, whereupon prices were cut. This reduced the revenues accruing to the concession-granting governments, and the coalition of nation states known as OPEC came into being in order to counter this effect.

1 J. M. Blair, *The Control of Oil*, Macmillan, 1977, pp. 159–166.
2 Edith T. Penrose, *The Large International Firm in Developing Countries*, Allen and Unwin, London, 1968, p. 150.

The United States: a Conservation Model? 33

The OPEC countries in their turn, however, followed the policy of meeting demand. In spite of an attempt by Venezuela to secure an agreement on the regulation of production, OPEC has limited its activities to pricing, and in practice 'the oil producers have [taken] into account the volume of demand in the consuming countries under normal circumstances.'[1]

It is apparent that there are loopholes in the United States' regulation of oil production. Individual States are not compelled to regulate; individual producers are not obliged to agree to the exploitation of a field as a unit; the rules vary from State to State. There is also freedom to differ from the official estimate of demand; and there are some States that choose not to attempt to match supply to demand. That regulation by the States can be loose is recognised by the Federal authorities which attempt, through legislation known as the Hot Oil Act, to prevent the passage to another State of oil produced in excess of the quota laid down in the State of origin. All these characteristics spring from the nature of the United States – the belief in individual freedom and the latitude enjoyed by the individual States.

The regulation of production in the United States, then, has contributed something to conservation, difficult though it is to compare what has happened with what might have happened. Robert E. Hardwicke, writing to E. W. Zimmermann at the end of 1954, opined 'that the ultimate recovery will be increased 50 per cent as the result of conservation practices', then added in a postscript: 'I really believe that ultimate recovery (including oil liquids from gas) as a result of conservation will be 100 per cent more than would be recovered without conservation practice.'[2]

It is conceivable that Mr Hardwicke exaggerated, for paradoxically the very policy of conservation also worked against conservation. It maintained the domestic price for oil

---

1 Yousef Sayegh, *The Social Cost of Oil Revenues*, First Arab Energy Conference, 4–8 March, 1979, Abu Dhabi, p.2.
2 Zimmermann, op, cit., p. 277.

above what it would otherwise have been and certainly above the price being charged in the 1950s and 1960s by the low-cost producers of the Middle East. 'So while the domestic price was being maintained by intensified production cutbacks in 1954, imports increased sharply and so-called voluntary controls (over imports) were instituted. When the same thing happened at the higher price level of 1957–58, while prices abroad were dropping under the impact of growing supplies in weakening hands, the pretence of voluntary controls had finally to be dropped and mandatory controls instituted.'[1]

Mandatory import quotas were introduced in 1959, and were to remain in operation for fourteen years – right up to the moment, that is, when OPEC made use of its power to raise world oil prices. 'By sharply limiting the use of foreign oil for fourteen years the import quota obviously hastened the depletion of domestic reserves though the extent is probably incalculable.'[2] Table 2 shows US production and trade of crude oil and petroleum products. By choosing to mulct her own reserves rather than countenance dependence on overseas supplies, the United States ultimately made her dependence on the outside world all the more sudden and all the greater, over 40 per cent of her requirements now being imported. It is only the rapid build-up in production from the Alaskan North Slope which has prevented the decline in total US production from being much more severe. It is the sudden dependence on the outside world of a once self-sufficient country which explains American psychology toward, for example, prices.

The general controls on prices introduced by President Nixon in 1971 survived for a time in the case of oil. Since 1975, and the hoist in OPEC prices, subsequent administrations have tried to beat a retreat. They have, however, been torn between conflicting considerations – the desire, on the one hand, to deny the domestic producer a

1 Alfred E. Kahn, quoted in John M. Blair, *The Control of Oil*, Pantheon, New York, 1976, p. 170.
2 Blair, op. cit., p. 186.

Table 2: *Production, exports and imports of crude petroleum and refined products (a)*

	Million barrels per day		
	Production of crude oil	Exports of crude oil and refined products	Imports of crude oil and refined products
1950	5.7	0.2	0.9
1955	7.3	0.3	1.3
1960	8.0	0.1	1.8
1965	9.0	0.1	2.5
1970	11.3	0.1	3.4
1971	11.2	0.1	3.9
1972	11.2	0.1	4.7
1973	10.9	0.1	6.3
1974	10.5	0.0	6.1
1975	10.0	0.0	6.1
1976	9.7	0.0	7.3
1977	9.9	0.0	8.8
1978	10.3	0.0	8.3
1979	10.2	na	na

na: not available.
(a) Includes natural gas liquids.
SOURCES: USA, Department of Energy; United Nations, *World Energy Supplies 1950-74* pp. 201, 296; *Petroleum Economist,* Volume XLVII, Number 2, February 1980, p. 87.

windfall profit which would result from too high a price and, on the other, the wish to discourage the imports which would follow if the domestic price were kept too far below the world price. And so they have fudged. One complex system has supervened upon another, still leaving the American consumer, however, paying a price below the world price. This anomaly cannot last, with its consequence of encouraging consumption and thus depletion of US oil. The day is not distant when the American consumer will have to adapt himself to his country's dependence on the outside world just as the Britisher is having to adapt himself to his country's loss of empire.

Meanwhile, is there anything that the United States can do to mitigate her dependence on the external world? Certainly she can undertake further exploration within her own territory – though certain parts of the eastern Rockies and Alaska have been debarred to exploration in order to preserve the wilderness – but the better hope probably lies in the use of secondary and advanced recovery techniques in already known reserves. Historically, oil has been produced with the help of the energy already existing in the reservoir. This is known as 'primary recovery'. 'Secondary recovery' is obtained through the injection of water or gas, or some combination of the two, after some degree of primary recovery. 'Enhanced recovery' takes place when further supplements to water and gas, including certain chemicals, are used – processes which are increasingly being applied to reservoirs early in their life, sometimes before either primary or secondary methods have reached their economic limit, so that the term 'enhanced recovery' has come to displace the expression 'tertiary recovery'.[1]

The combined use of primary and secondary recovery can produce between 20 and 50 per cent of the oil originally in place. How much extra can be produced with enhanced recovery is as yet unclear. Certainly secondary recovery has spread. Between 1950 and 1970 primary production in the United States remained roughly constant at 5 mbd, while secondary production increased rapidly from about 1 mbd to 3.5 mbd. It is estimated that in 1976 about 45 per cent of the total US daily production was obtained from fields in which secondary recovery operations had been applied. In contrast, production from enhanced oil recovery operations is estimated in 1975 to have been only 350,000 barrels a day. This meagre result from enhanced recovery operations is due to the long lead time, the extent to which the geology has to be thoroughly explored, and the cost, estimated to be at least $20 a barrel. 'From the results, it is obvious that enhanced recovery operations, when operated under the most favour-

---

1 *Reservoir Engineering*, op. cit., p. 252.

The United States: a Conservation Model? 37

able conditions, cannot compensate for the declining productivity of known fields in the USA.'[1]

The United States has some lessons on conservation to impart to other oil-producers. It is clear that the avoidance of waste requires the unified development of a single field, which not only needs action on matters such as well-spacing, but in the Middle East also implies negotiation and co-operation between countries over shared oilfields. As the first oil producer the US was the pioneer in physical controls of production to prevent waste of oil. The nature of oil's formation in the ground makes it important that production is controlled to prevent irrecoverable waste. 'Prorationing' has had the effect of conserving oil, and the importance of conservation is being increasingly learnt in other oil-producing countries. It also caused prices to be higher for a period in the US than they would have otherwise been.

However, in terms of a coherent depletion policy there seems little to be learnt from the US. For, while the US had a period of conservation, this was both preceded and followed by a degree of dissipation. This dissipation was caused in the first case by excessive competition among producers, leading to waste; then by the discouragement of imports of cheap Middle Eastern oil; and finally by a low domestic price for oil which encouraged consumption and the excessive depletion of US oil reserves. If oil is to be conserved in the ground against future eventualities, a wider policy on the part of the government than that hitherto prevailing in the United States is required. Meanwhile, in spite of active exploration in the United States and experiments with enhanced recovery, the ratio of reserves to output (i.e. consumption) continues to decline and the dependence of the United States on the outside world to increase, which enables other oil-producing countries to raise their prices without necessarily having to cut their production.

1 Ibid. p. 281.

CHAPTER 3

Canada: the Federation under Strain

Canadian policy on oil depletion results from the country's geographical relationship with the United States, the fact that some 95 per cent of the oil is extracted by vertically integrated firms, mainly from the United States (both these factors compel some assimilation to the United States) and the federal nature of the country's constitution.

Although oil was first discovered in Canada as long ago as the middle of the nineteenth century, it is only since the Second World War that extraction has assumed any importance. Table 3 shows that even in 1950 crude oil production was only 0.1 mbd. Jurisdiction over oil production was delegated in the main to provincial governments. Unlike their counterparts in the United States, these were not encumbered by the automatic attachment of mineral rights to property rights, so they were not confronted with quite the same problems concerning conservation as the United States, though the answers turned out to be similar.

The practice has been to sell blocks of land to the highest bidder. This could be called 'free enterprise' practice. It can, however, give rise to problems – for example, where one oil reservoir is covered by two or more blocks. In such a case the provincial government has had to introduce regulations to ensure the proper spacing of wells and to encourage 'unitisation' or the operation of the field as a unit. Indeed the relevant Minister may make 'unitisation' compulsory, a practice which goes further than that of the United States. The problem of a spread of fields over more than one province does not appear so far to have arisen, though its occurrence is not, of course, impossible. In addition, output has been controlled to match demand as foreseen. The exploitation of oil in Canada was not marked by the initial waste which characterised the United States (except for the case of the Turner Valley field, which was in production in the 1920s and 1930s before conservation laws were passed in Canada), but otherwise Canadian provincial practice has followed the model of the United States.

Table 3: *Production, exports and imports of crude petroleum and refined products (a)*

	Million barrels per day		
	Crude oil production	Imports of crude petroleum and refined products	Exports of crude petroleum and refined products
1950	0.1	0.3	0
1955	0.4	0.3	0.04
1960	0.5	0.4	0.1
1965	0.9	0.6	0.3
1970	1.4	0.8	0.7
1971	1.5	0.8	0.8
1972	1.7	0.9	1.0
1973	2.0	1.0	1.2
1974	1.9	0.9	1.0
1975	1.7	0.9	0.8
1976	1.6	0.8	0.5
1977	1.6	0.7	0.4
1978	1.6	0.7	0.4
1979	1.8	na	na

na: not available
(a) Includes natural gas liquids.
SOURCES: UN, *World Energy Supplies 1950–1974*, pp. 201, 295; UN, *World Energy Supplies 1973–1978*, pp. 128, 161; *Petroleum Economist*, Volume XLVII, No. 3, March 1980, p. 135.

The model was, however, inadequate. This inadequacy became apparent with the approach of the fateful years 1973 and 1974, when the members of OPEC raised the price of crude oil nearly fivefold. Not that the price increase had been without presage. There had been oil 'crises' in 1956, when the Egyptian Government nationalised the Suez Canal, and in 1967, when Israel defeated Egypt and Syria in a devastatingly brief war. Until 1973 the price of OPEC oil had been below that of United States oil, the United States Administration having in 1959 disallowed international prices within its own territory and introduced restrictions on imports. At that stage world prices were below the United States prices. From 1973 onwards, however, the international price rose far above the United States price. And in that year, although before the OPEC decision, Canada changed its policy.

Canadian oil production is at present confined to the Western provinces, principally Alberta. It is true that there are promising geological signs of oil off the Eastern coast and in the Arctic, but the promises have yet to be fulfilled. By 1979, Alberta was providing more than 80 per cent of Canada's domestic oil and gas production. However, since about 1974 Alberta's oil fields have been operating at capacity, and their production has already begun to decline. Alberta's conventional oil reserves, like those of any other country or province, are limited, and indeed there has been only one significant oil discovery in the past fourteen years. Table 3 shows how production has fallen from its peak in 1973, although there was some recovery in 1979.

According to a policy laid down in 1961, the Western provinces were to confine their Canadian sales to the areas west of the Ottawa valley, and sell the surplus to the United States, while the Eastern provinces could import oil from overseas. It has been seen that up to 1973 the United States price was above the world price. Canadian Western producers, therefore, were enjoying the higher United States price, while consumers in East Canada were buying at a lower world price. With world prices shooting above the United States price in 1973, everything was changed. The Eastern provinces were now to be disadvantaged relatively to the Western, and even the Western provinces would lose if the international price continued to rise above the United

States price while their sales were confined to the United States. The impending change in circumstances indicated a change in policy.

The more important part of that change was the reduction in Canada's oil exports to the United States, This began in March 1973, several months before OPEC decreed an increase in price, with the introduction of controls on the export of crude oil. After the OPEC increase, the Canadian National Energy Board, a quasi-judicial body which recommends whether or not a pipeline or an export permit be granted, went a step further, and advised that exports to the United States be phased out. Although the Federal Government is not obliged to accept, the recommendation was accepted and exports to the United States fell steadily, being more than halved between 1973 and 1976 with net exports of light crude oil being almost all phased out by the end of 1979, although oil exchanges to the amount of some 100,000 barrels a day were maintained. This meant in effect that, as far as oil was concerned, Canada was pursuing a policy of 'self-reliance' – that is, a policy of relying on her own resources and reducing to a minimum her dependence on the outside world. The prime reason for the cutback in crude oil exports to the United States was a decline in light crude oil reserves which emphasised the importance of conserving domestic resources for domestic needs and reducing the country's vulnerability to oil imports. The reduction in exports as well as in total production is brought out in Table 3, and reflects the declining production and level of reserves of conventional oil in Alberta.

The cut in exports has been supplemented by an increased emphasis on enhanced recovery. Alberta is experimenting with many different methods, and now has from any one field an average of oil which is high by international standards.

The reduction and, *a fortiori*, the stoppage of exports to the United States was accepted by Alberta in spite of the considerable political influence which, through oil, she had acquired in the Canadian Federation. On the other hand, she successfully resisted attempts by the Federal Government to introduce tax measures for a sharing of 'windfall' profits resulting from the greatly increased world prices which theoretically could now be charged in Canada.

In fact, the enhanced world price was not charged. Instead, there was instituted throughout Canada a uniform price for oil in place of the previous two-tier system – the United States price in the West and the lower international price in the East. The uniform price meant a higher price to the East, because of rising import prices, the increase being financed in part by a levy on such exports as were still sent by the West to the United States, in part by an excise duty on all gasoline. Had the new uniform price been equivalent to the international price as enhanced by OPEC, the Western provinces could have regarded themselves as more than reasonably compensated. The single price was, however, deliberately kept below the enhanced international price in order to ease for Canadian industry the transition to dearer energy.

This was still the case in January 1980, when the Canadian price was raised to $14.75 a barrel, compared with a world price of over $30 a barrel. Canadian refineries received a Federal Government subsidy of about $17 a barrel where necessary, to prevent the domestic price of refined oil from rising further. The Canadian election of February 1980 was fought partly on the issue of whether or not domestic oil prices should be raised nearer the international level, and the electorate clearly said 'No'. The rationale of Canadian policy as traditionally pursued was justified to the extent that Canada was less affected by the world recession of 1974 than other 'developed' countries. A price was paid, however, in the shape of a heavily aggrieved Alberta, which has become increasingly frustrated at the failure to raise prices even to the levels fixed in the United States.

Whether the single price encouraged the conservation of supply is more doubtful. It could be argued that the maintenance of a price below the full international price gave an unnecessary boost to consumption. There is some confirmation of this in a statement made by the Canadian Prime

Minister in April 1975: '. . . my colleagues in the Government and I have come reluctantly to believe that the price of oil in Canada must go up – up towards the world price. It need not go all the way up. We should watch what happens to the world price and decide from year to year what we should do.'[1]

It is possible that any increased consumption resulting from the fact that the price has 'not gone all the way up' could be met, in part, by 'non-conventional' oil from the vast Athabasca oil sands of Alberta. The oil in these sands either consists of bitumen or borders on bitumen. Unlike 'conventional' oil it cannot be pumped to the surface but has to be mined or converted into a fluid through steam almost continuously applied. Either process is extremely costly. In recognition of this fact, projects to exploit the oil sands are given the incentive of international prices; the consumer would still be charged a relatively low price, while the producer would be subsidised through a fund based on a levy on all crude oil running through Canadian refineries.

It has, however, hitherto been the policy of the Canadian Government to keep shut in the ground oil the cost of extracting which is above the domestic price. For example, the Syncrude project, designed to convert bitumen into crude oil, although it has been in operation since 1977 and produces 125,000 barrels a day, is not yet a financial success. It is for this reason that the Alberta sands have remained largely untapped. The ultimate logic, however, of the doctrine of self-reliance could mean, if imports continued to rise – and it is estimated that they could amount to just under 50 per cent of Canadian demand by 1985 – the extraction of domestic oil at a cost equal to, or even above, the international price. It is in support of the objective of self-reliance that oil from the sands is priced (as far as the producer is concerned) at the international level.

In the light of the prospect of increased imports, the concept of self-reliance has now given way to that of 'self-

1 *An Energy Strategy for Canada,* Ministry of Energy, Mines and Resources, Ottawa, 1976, p. 21.

sufficiency'. There is a tenuous distinction between the two; 'self-sufficiency' means a lower Canadian supply than 'self-reliance' and is to be obtained through greater conservation on the part of the consumer and through the substitution for imported crude oil of Canadian natural gas and other forms of domestic energy.

Gas, like oil, is found mainly in Western Canada and – again like oil – is subject to the jurisdiction of the provincial governments. Unlike oil, however, the rate of discovery has exceeded the rate of growth in domestic consumption; and, by traditional policy, what is in excess of foreseen home requirements could be exported. The natural export market was the United States. Unfortunately for Canada, in the 1950s the United States Federal Power Commission was required by the Supreme Court to regulate the price of gas – at a low level. As a result the price of gas in the Eastern cities of Canada was lower than the thermal equivalent in oil. The depletion of the Western gas reserves was thereby accelerated and that of the oil reserves – through an accident – retarded. Late in 1975, it was decided to raise the price of gas but, as in the case of oil, 'in a manner that affords an opportunity for Canadian consumers to adjust to higher prices.'[1] This decision was not reached without a protracted and contentious dialogue centred on oil pricing between the Federal Government and the increasingly challenging Government of Alberta. Such are the politics of a federal constitution in which one of the member provinces becomes mighty through gas or oil.

In broad summary, although Canada, because of her climate and her distances, probably consumes more energy per head than any other country in the world, she is almost a paragon in depletion policy. The earlier post-war years were marked by a prudent exploitation of the oil resource, and the world price increase of 1973/74 was anticipated by a curtailment of

1 Ibid., p. 23

exports to the United States, the curtailment implying the conservation of more in the ground, particularly of the oil in the sands. This action was supplemented by an increased emphasis on enhanced recovery. It is true that conservation in this broad sense was to some extent offset by the extra consumption induced by a domestic price which was lower than the world price, but it is doubtful whether it was wholly offset. Canada's real problem lies in the future. She cannot rely for long on present conventional sources of either oil or gas, conventional oil having a life of little more than ten years and conventional gas a little less than thirty years, as can be seen from Chart 1. Her future, therefore, rests on exploiting new sources, and new sources present her with a dilemma.

She wishes both to be self-sufficient – that is, to import as little as possible – and to alleviate the effect on her citizens of a rising world price. Can she have it both ways? The rising world price will certainly accelerate the exploitation of the Alberta sands as well as the exploration of the inhospitable North, both processes probably entailing new technologies. As a result, Canada could well become almost, if not entirely, self-sufficient in oil. This is the objective for 1990. In that case the internal price might be held fractionally below the escalating world price. If, however, the attainment of self-sufficiency is frustrated, either Canada will have to import at the world price or else the restive Alberta will insist on exporting at the world price, in which latter case the Federation would be placed under further strain. The movement of the domestic price up towards the international level would then become inevitable. At the moment the producers of oil from the Albertan sands can enjoy the full international price, but the users are subsidised by industry, the Alberta Government, and the rest of Canada through the Federal Government. How long this arrangement can last is a crucial question for the continued political unity of Canada. Under the fiscal policy existing early in 1980, over the next few years Alberta would receive more than ten times as much revenue from oil as would the Federal Government in Ottawa. Such a division of resources could clearly create difficulties for the Federation, a requirement of which is a reasonable equality of resources between its constituent parts.

Chart 1: Life indices of conventional crude oil and marketable gas

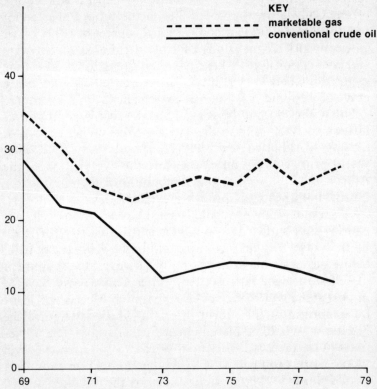

SOURCE: *Conservation in Alberta,* Energy Resources Conservation Board, Calgary, 1979.

There is one other aspect of Canadian oil policy which has implications stretching beyond Canada. In 1975 there was established a national oil company under the name of Petro-Canada, with powers to take part in all stages of the industry's operations, from exploration to marketing. It was not established because the privately-owned companies were reluctant to explore, which they are not, but chiefly because the Government needed to have a better knowledge of the oil and gas resource potential of the country, and it was felt that a national oil company would direct its principal efforts to obtaining this knowledge. It was also recognised that in arrangements with the other oil-producing countries of the world for the purchase of oil, a country's opportunities of acquiring secure supplies could be greatly enhanced if it could deal through its own national oil company with other government companies as well as with other private producers. And it is government companies that now dominate the scene. Further, the Canadian Government wished to acquire a more extensive knowledge of the industry so as to enable it better to judge how much to tax. Oil companies are suspected of enjoying an economic rent – that is, a surplus of revenue over costs, including in costs a 'normal' profit plus an element to compensate for the risk in exploring and developing; the profit allowable for risk is clearly arbitrary. It is the profit above that which is difficult to assess, but which governments seek to tax. Subsequent chapters will show how almost all oil-producing countries, with the exception of the United States, have felt the need for a national oil company to help the government capture the economic rent.

Canada, like the United Kingdom, is showing a steady increase in the import of manufactured goods and a steady decrease in their export; in other words, she risks, like the United Kingdom, 'de-industrialisation'. Her oil companies have their origin in the United States, undertaking their research and development there and placing there their orders for technologically more advanced products. If, through her own oil company, Canada could undertake such work within her own borders, this could help stem the process of de-industrialisation.

Now the establishment of manufacturing, stemming from the sale of oil, has been in the minds of some of the oil-producing countries of the Middle East and of South America. They see in oil the key to industrialisation. Canada, however, has not seen it in this way. It is true that one-third of the Albertan revenues from hydrocarbons is allocated to a fund known as the Alberta Heritage Savings Trust Fund, which has generated much interest elsewhere and has produced some supporters for a similar idea in the United Kingdom. Alberta's revenues from oil have risen to $3,500 million in 1979, and may total as much as $50,000 million in the next five years. The Trust Fund now stands at about $6,000 million, and will be increasing at over $1,000 million a year. There are thus substantial resources available for Alberta to attempt to industrialise. Before oil became important, agriculture was Alberta's main industry, and it still remains the second most important, with production in 1978 valued at $2,300 million, compared with $9,200 million for the oil and gas industry. However, of the monies flowing to the Fund only 20 per cent may be invested in capital projects, and few of these projects can be described as industrial. True, there is an investment in oil sands technology research and in the Syncrude project, and there has also been substantial investment in petro-chemical expansion. But other investments are in airports, farming, the environment, and medical research, while loans have been made to other Canadian provinces.

And yet Alberta is anxious to industrialise. It is simply that oil and gas revenues have not been used for the purpose. Two-thirds of Albertan oil revenues are absorbed into the general account, and Alberta certainly has higher growth and lower unemployment than elsewhere in Canada, symbolised by a gigantic construction boom. Yet this of itself does not spell industrialisation, even in a semi-industrialised country. Can oil lead to industrialisation in countries which have no industry at all? And if it does not, what is the effect on the rate of depletion? Will a failure to industrialise quickly cause, as in Canada, a reduction in exports and prompt a policy of stretching the oil over time? The chapters that follow will seek to answer these questions, for they apply to more than one country. Meanwhile, let it just be noted that Canada's reaction to the oil 'crisis' has been simple – a reduction in exports.

United States and Canada

Blair, J. M., *The Control of Oil*, Macmillan, 1977.
Energy Resources Conservation Board, *Conservation in Alberta*, Calgary, 1979.
Macdonald, Stephen L., *Petroleum Conservation in the United States*, Johns Hopkins University Press, Baltimore, 1971.
Ministry of Energy, Mines and Resources, *An Energy Strategy for Canada*, Ottawa, 1976.
Mughraby, Muhamad A., *Permanent Sovereignty over Oil Resources*, Middle East Research and Publishing Centre, Beirut, 1966.
Penrose, Edith T., *The Large International Firm in Developing Countries*, Allen and Unwin, London, 1968.
Sayegh, Yousef, *The Social Cost of Oil Revenues*, First Arab Energy Conference, Abu Dhabi, 4–8 March 1979.
Zimmermann, Erich W., *Conservation in the Production of Petroleum*, Yale University Press, 1959.

CHAPTER 4

Latin America: Sowing the Oil

There are three main oil-producing countries in Latin America – Venezuela, Mexico (the first country ever to nationalise oil back in 1938), and Ecuador. Brazil used to be a producer, but has now ceased to be to any significant extent.

Venezuela, like Iran, produced oil before she had any kind of industry, and, like Iran, has deliberately used oil in an attempt to industrialise, though her approach to the problem has been more cautious, to start with at any rate. Having thrown off the yoke of Spanish dominion in 1811, she saw herself, in later decades, as the victim of a new colonial power – the multinational oil companies. These began to appear on the scene towards the end of the First World War, obtaining concessions from the Venezuelan Government, the owners of all subsoil resources. Subsequent Venezuelan administrations, alternating between military dictatorships and elected governments, have been concerned with a double objective: to obtain for the State a greater proportion of the oil revenues and a greater control over oil policy. Throughout, however, there has been some concern with conservation in the ground, though the degree of this concern may have varied from regime to regime.

For example, a law of 1917 fixed a maximum duration for concessions of thirty years. Any concession which was not exploited in the first three years after it had been granted was to be returned to the Venezuelan Government, and half of each production area was to be set aside as a national reserve, the property of the State. The law of 1917 was clearly an attempt to establish some control by the State. Then, with a change of political regime, the pendulum swung in favour of the companies. A few years later, in 1921 and 1922, the law was changed to give longer terms for all concessions and to exempt the oil companies from paying duties on imports. It was calculated by Torres, the author of the law of 1917, that in the seven years leading up to 1930 exemptions from customs duties had reached a total value of 219 million bolivares, while only 187 million bolivares had been collected by the Government in taxes from the oil companies. 'The companies take our oil and the Government pays them to take it away.' Put more brutally, the oil companies were seen as 'without regard to the local economy and in the absence of policies to protect Venezuelan nationals or their interests.'[1] Whether the oil companies could have done more for the local economy can perhaps be more objectively judged when it is seen what nationalised companies have done.

Possibly even more important than extra revenues for the State was the winning of a 'significant share in decisions on the expansion and policies of the companies in general, and above all in relation to the new areas they will be assigned for exploration and exploitation.'[2] This larger objective was clearly linked, in part at any rate, with conservation in the ground. The objective was attained by successive steps from, roughly, the end of the Second World War onwards. The 1945 Junta enacted policies to ensure that the Government received 50 per cent of oil company profits. A law of 1948 'provided for a 50 per cent tax on any sum by which a company's net profits for any year exceeded the Government's total revenue from that company's activities in Venezuela.'[3] This formula subsequently spread to Iran in 1949, Saudi Arabia in 1950, Kuwait in 1951, and Bahrain and Iraq in 1952.

A culminating step was the Hydrocarbons Law in 1943, under which all oil was brought within one legal code. The

1 Loring Allen, *Venuezuelan Economic Development: a Politico-Economic Analysis*, JAI Press, Connecticut, 1977, p. 37.
2 Romulo Bentancourt, *Venezuela's Oil*, Allen and Unwin, London 1978, pp. 90, 91.
3 Franklin Tugwell, *The Politics of Oil in Vnezuela*, Stanford University Press, 1975, p. 45.

most important result of this single code was that, directly or indirectly, it terminated the concessions of all companies by around 1980. The Government was later empowered to collect technical information from the companies through the Coordinating Commission on Conservation and Marketing of Hydrocarbons, which could monitor the techniques used for extracting, selling and manufacturing by-products. Whether the Commission was fully equipped to carry out this task may be questioned. Finally, the oil pipeline became a common carrier, instead of being the private possession of the oil companies. There followed in 1960 the establishment of a national oil company, the Corporación Venezolana del Petroleo.

Throughout the 1960s there was a continual struggle between the Government and the oil companies, with the Government gradually gaining the upper hand. This strategy entailed great risks for Venezuela, as the companies reduced investment and production, and it was only in the wake of the 1973/74 oil price rise and the changed state of the world energy balance that Venezuela finally triumphed over the foreign oil companies. The evolution ended in the Organic Law of 1975, under which all the oil companies were to be nationalised as from January 1, 1976. Henceforth, there would be no more concessions, only contracts for technical assistance, signed with the national oil company, Petroleos de Venezuela.

The outcome of the struggle between State and oil companies was certainly to reverse the proportions of the oil revenues taken by the companies and the State, as is indicated in Table 4.

While the Government's search for an increasing share of the oil revenues was taking place, the Venezuelan economy continued to expand. For example, between 1936 and 1958 the gross national product per head grew at the rate of 7 per cent a year, to become the highest in Latin America. Oil was clearly contributing to this expansion. The oil companies, however, could see the writing on the wall and were bound to react to increasing Government penetration into the field

Table 4: *Government and company shares of oil revenues*

	Percent of total oil revenues			
	1950–57	1958–64	1965–69	1970–73
Companies' share	47	33	32	19
State's share	53	67	68	81

SOURCE: F. Tugwell, op. cit., pp. 179–181.

of oil. Their reaction took the form of declining investment. Exploratory wells drilled fell in number from 598 in 1958 to 170 in 1963, and again to 75 in 1967. Investment all round fell from a peak of 1,822 million bolivares in 1957 to 474 million bolivares in 1962.

As a result, when nationalisation took effect on January 1, 1976, the Government announced a production target of 2.2 to 2.3 mbd, as against the 3.7 mbd produced in 1970, the highest level Venezuelan production had reached. This reduced target was put forward in the name of conservation, though there were those who contended that even this reduced figure was too high. In reality, because of the preceding decline in investment, it was probably as much as the industry could produce. The target of 2.2 mbd later became embodied in the Fifth Plan (1976–1980), though the Government also aimed to raise capacity to 2.7 to 2.8 mbd. This figure is unlikely to be reached in the near future, the highest production likely in 1982 being around 2.4 mbd, the same production that was reached in 1979. Table 5 shows how production and exports have fallen steadily since 1973.

In the distance, it is true, there is the oil of the Orinoco oil belt, a liquid oil, unlike the solid bitumen of the Albertan oil sands, but also a dense oil, necessitating upgrading in refineries. Nobody knows how much oil the Orinoco belt contains, and any figure put forward at this juncture would be entirely conjectural. However, the sums required for investment to develop the Orinoco are so substantial, and the time lag before significant production would be so long that it would be unrealistic to expect much production from this source for at least a decade, and perhaps much longer. A production of around 1 mbd might perhaps be expected by 1990. Venezuela's difficulty in raising production, in spite of a trebling in recent years in expenditure on exploration, will inevitably be reflected in her ability to export.

Table 5: *Venezuelan oil production and exports*

	Million barrels per day		
	Crude oil production	*Exports of crude oil*	*Exports of refined products*
1950	1.5	1.3	0.2
1960	2.8	2.1	0.7
1965	3.5	2.4	0.9
1970	3.7	2.4	1.0
1971	3.5	2.3	1.0
1972	3.2	2.1	0.9
1973	3.4	2.1	1.0
1974	3.0	1.8	1.0
1975	2.3	1.5	0.6
1976	2.3	1.4	0.8
1977	2.2	1.3	0.6
1978	2.2	1.2	0.7
1979	2.4	na	na
1982(a)	2.2 to 2.4	1.5 to 1.6	

na: not available.
(a) Author's forecast.
SOURCES: OPEC, *Annual Statistical Bulletin 1977*, pp. 29, 30, 75, 80, 81; OPEC, *Annual Report 1978*, pp. vi, xiii; *Petroleum Economist*, Volume XLVII, Number 3, March 1980, p. 135; United Nations *World Energy Supplies 1950–74*, pp. 209, 308; United Nations *World Energy Supplies 1973–78*, p. 165.

The other determinant of the level of exports is the rate of domestic consumption. The price charged to the domestic consumer is one of the lowest in the world, and is well below the cost of production to the producer – $0.5 per gallon of gasoline in 1977 as against $1.5 in the United Kingdom. In developing countries motoring generally increases more rapidly than the gross domestic product, and the Venezuelan internal market for oil, particularly gasoline, has recently been rising at a rate of over 12 per cent a year. This rate of increase might fall in the 1980s to some 7 per cent. Equally, it might accelerate to around 14 per cent.

There is no precedent for a sound prediction. However, an internal market of about 0.4 to 0.6 mbd in 1982 is a reasonable estimate, given consumption of 0.25 mbd in 1977.

The possible scenarios, then, for Venezuelan production in the early '80s are shown in Chart 2. The dashed line 1 represents the capacity of 2.7 to 2.8 mbd aimed at by the Government. Line 2 is maximum feasible production, rising from its 1977 level of 2.2 mbd to 2.4 mbd in 1979. Line 4 shows how the internal market for oil is expected to develop, leaving the difference between production and consumption available for export. This is represented by line 3, and even on the most optimistic assumptions oil exports are expected to fall between 1977 and 1982. If the more pessimistic assumptions underlying Chart 2 were extrapolated, a time could be reached when Venezuela would cease to be an oil exporter, although it is possible that by then the Orinoco belt and offshore sources will have become significant producers.

From the point of view of the rate of growth of the gross national product in Venezuela the price of oil matters as much as the output. It also matters for the Government's revenues, for the nationalised company pays to the Government in royalty and tax 72 per cent of its net income. Indeed, both government revenues and the economy generally are heavily dependent on oil and the returns to the oil industry. During the 1970s up to 90 per cent of government revenue and 97 per cent of export earnings were derived from oil. Venezuela's dependence on oil for her development and industrialisation goes back to before the beginning of the Second World War, but their interdependence has increased in recent years. With the increase in the price of oil in 1973–74 the contribution of oil to the annual rate of growth of the Venezuelan gross domestic product was 25 per cent. This contribution, as well as the rate of increase in the gross domestic product as a whole, fell sharply with the fall in the real price of oil in 1975, only to rise again with subsequent increases in the price of oil.

Chart 2: Projected production, exports and domestic consumption of oil 1977 to 1982

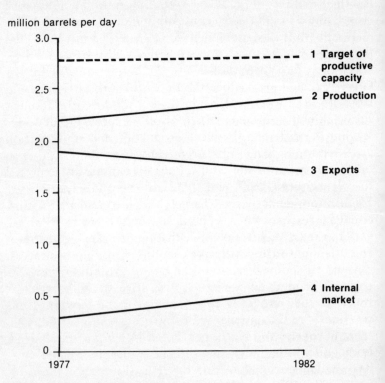

SOURCE: Author's estimates

It is clear that a government so dependent on oil, both output and price, for its gross domestic product and faced with increasing difficulties in extracting oil, will be tempted to push for a higher price, for only thus could its rate of growth be maintained.

Though there is still controversy surrounding conservation in Venezuela, it is unreal. Ultimately any government's policy towards conservation is determined by the social forces by which it feels compelled. The two sides of the controversy in Venezuela have been well put by Mr Arthur Lewis: 'Politics in contemporary Venezuela centres in a power struggle between two major groups of forces. One represents a traditional order – the land barons, the politically inclined military officers, the large merchants, the conservative clergy – all those elite elements that dominated the republic from the beginnings of the nationhood to the end of the Second World War. They represent the status quo. Their outlook is static. They resist change. The other represents the masses of the people – the white collar workers, organised labour, the landless peasantry – all those popular elements aspiring to a better life. This group is a revolutionary force. It proposes fundamental reform. It demands popular rule, redistribution of wealth, new clan alignments. Thus the political struggle between the two involves far-reaching economic and social issues. Venezuela today is in the throes of revolutionary transformation. The upheaval is still in the early stages.'[1]

The struggle thus described is epitomised by the contrast between the skyscrapers and Cadillacs of Caracas and the shanty villages that rim the red surrounding hills. And in a country where tax charge on an income of $200,000 is only twenty per cent, the pressure of the masses, and therefore, the electoral pressure on a government, will make for the maximum extraction of oil. In other words gross inequality in income will force, certainly under an elected regime and possibly under a dictatorial regime, a rapid depletion of the oil resource.

Venezuela, it is believed, is the country where the phrase 'sowing of the oil' originated. 'The country has not possessed a few clear brains able to organise from a position of power what has been called the sowing of the oil: the conversion of this wealth of exploitation money into national industries

[1] Quoted in Edwin Lieuwen, *Venezuela*, Oxford University Press, 1967, p. 156

which might strike permanent root.'¹ A developing country, seeking to industrialise itself, has normally begun by erecting a protective barrier behind which the country has been able to produce elementary consumption goods – textiles, for example. The problem has been to advance beyond this stage to the manufacture of intermediate and capital goods, an advance requiring a sufficiency of liquid capital. The oil-producing countries have long been fortunate in being free from the usually fatal constraints of inadequate capital. But if oil revenues have provided the liquid capital they have not made more advanced industrialisation automatic. Venezuela, it is believed, was the first country to seek to use for this purpose the liquid capital derived from oil. It used the power in several ways.

First, it selected a specific region – Guayana – in the south-eastern part of the country as a centre of regional development. The reason for the selection of this area was the combined abundance of iron ore and water power. '. . . the discovery and exploitation of large iron ore deposits in the Guayana region in the early 1930s by Bethlehem Steel, and the far larger discoveries in 1945 and 1946 by US Steel, as well as the series of optimistic reports by various geological expeditions and the history of gold and diamond mining in the region, all served to keep alive the myth of the Guayana as the storebox of Venezuelan treasures.'² The agency appointed for the development of the area was the CVG (Corporación Venezolana de Guayana). The creation of this agency was an imaginative enough stroke, while its endurance through alterations of government was, perhaps, even more remarkable. This endurance was due to the character of its head, General Alfonzo Ravard, a man of firm gentleness who steadfastly stayed aloof from politics and kept his eye fixed on the technological development of the area. Today the steel industry of Guayana shows steadily rising production, while the hydro-electric power industry supplies energy

1 Guillermo Moron, *A History of Venezuela,* Allen & Unwin, London, 1964, p. 152.
2 John R. Dinkelspiel, *Technology and Tradition: Regional and Urban Development in the Guayana,* p. 2.

not only to the steel industry, but also to the aluminium industry.

Secondly, while the Government receives 72 per cent of the net revenues of the oil industry, 40 per cent of these receipts are placed in an Investment Fund, which in turn feeds agriculture, residential construction and private sector industry. It is of course inevitable that when the Government is so involved in the economy, and receives so much of the income of the oil industry, it should play a major role in financing and encouraging private industry. There are several government financial institutions concerned, from the Venezuelan Development Corporation founded in 1946 to the Investment Fund set up in 1974. In this role the Government is aided by the country's presence in the Andean Pact[1], which endeavours to rationalise manufacture among its members. The most striking example of the advantage to Venezuela is that, while Venezuela might furnish some 20 per cent of the market for an Andean car, the country has been allotted the task of constructing the greater part of the engine, representing, say, some 45 per cent of the total cost. This fact suggests that Venezuela has what other developing countries lack – the beginnings of an engineering industry.

Finally, the State enterprises – namely, oil, steel, aluminium – so place their procurement orders as to promote within industry a more advanced technology. For instance, the oil industry places 40 per cent of its orders for valves, etc. with the domestic private industry, providing the firm with a foreign advisor if the firm itself is not up to the job. If the materials manufactured domestically are not competitive with those obtainable from abroad, the order is placed overseas. Thus the customary practice of developing local industry behind a protective screen is circumvented.

All in all, total public sector expenditure in Venezuela – that is, both of the Government and its main industrial satellites – represents some 65 per cent of the gross domestic product, while the ratio of investment to the gross domestic

---

1 The Andean Pact comprises Bolivia, Colombia, Ecuador, Peru and Venezuela.

product is over 35 per cent, as against under 20 per cent in the United Kingdom. This strenuous effort, promising though it is, has not been conducted entirely with reason. The Fifth Plan consisted of a list of projects, thought initially to match the financial resources likely to be available, but in the end vastly exceeding them and compelling resort, in spite of oil revenues, to borrowing from abroad. The approach to the Sixth Plan appeared to be more thoughtful, in that it aimed at knitting to each other related industries – that is, it aimed at both forward and backward integration; the details, however, had yet to be worked out in 1980.

The summary impression of the Venezuelan effort to sow the oil is that something is being achieved – aluminium parts, for example, are being exported to the United States. There is, however, one fatal flaw to the entire industrial effort: it is dependent on a price for oil which is well below the world level. It is not easy to foresee a Venezuelan government abolishing the subsidised price for domestic oil; those shanty villages have more power than their down-at-heel appearance might suggest. Behind the screen of subsidy Venezuelan industry can continue to develop for a long time to come. Without that screen it is entirely uncompetitive. It may well be that in time Venezuelan industry can develop to the point of being able to dispense with the subsidy; Venezuelan industry will then indeed have made it. Meanwhile it is being in effect financed by the Venezuelan oil industry.

Indeed, Venezuela's cheap domestic oil to finance her industrialisation is typical of all oil-producing developing countries. It is considered an essential stimulus to industrialisation. On the other hand, the validity of the policy has often been challenged on the ground that the charging of a world price for energy would force a country towards industries competitive with those of the outside world. Whatever the practical truth, one's judgement must be that Venezuela has a chance of sowing the oil, but only just a chance. That chance depends at present on charging the domestic private consumer and domestic industry a lower than world price for oil. The low domestic price in turn makes for a high rate of consumption and consequently of depletion, and must impair the ability to export oil. Unless, therefore, the Orinoco, with its large requirement for investment, can raise the general level of output, Venezuela is likely for some years to come to be producing up to her physical capacity.

An analysis not dissimilar from that of Venezuela can be applied to Mexico, where newly discovered reserves of oil are, as usual, much exaggerated. Mexico has, for the time being, a regime which gives absolute power to an elected President. It is in effect a party dictatorship, though an elected one. The fact of election exposes it to popular pressure. And Mexico is a country marked by extreme inequalities – between one part of agriculture and another, between one industry and another – with the result that, whereas during the decade 1968–77 the average income per head increased by some 18 per cent, the real income per head of the poorest Mexicans remained practically unchanged. It is inevitable that this degree of inequality will have its effect on the rate at which recently discovered oil is depleted. At present the officially declared policy is one of conservative exploitation. The outcome will, however, depend on the degree of pressure to which the authorities feel themselves subject.

A low rate of depletion would imply that the Government would have to increase taxes in general – as in Venezuela – and indeed increase the internal price of oil – low at present – to make good the revenues otherwise foregone. A more rapid rate of depletion would mean that the Government would remain dependent for a major part of its revenue on oil and would not need to raise the domestic price of oil. It would have adverse effects on agriculture, and the increased oil exports made possible through increased production would allow increased manufactured imports – to the dismay of the manufacturing industry which, in Mexico, preceded oil. There is a conflict, not confined to Mexico, between, on the one hand, a fast rate of increase in oil production with an accompanying lower rate of growth in manufacturing industry, and, on the other, a slower rate of oil extraction and a faster growth in manufacturing.

Table 6 gives details of Mexico's production of and trade in crude oil and refined products since 1950. It is clear that the second half of the 1970s saw a rapid build-up in production and exports. The origins of the nationalised oil company – Pemex – go back to 1938, when the Government decided to over-rule an inadequate wage offer by the companies. Pemex has ever since remained proudly nationalist, with an impressive research establishment staffed entirely by Mexicans. Pemex is aiming at a possible output target in 1982 of 4.0 mbd, but a more realistic estimate might be 2.75 mbd, since there are likely to be delays in construction. Internal consumption is at present growing at 8 to 9 per cent a year – a rate of growth which appears unrealistically low in the light of Venezuelan experience – which would imply holding exports steady at 1.1 to 1.4 mbd. It may be technically possible for Mexico to be producing 8.0 mbd in ten to fifteen years, if the reserves in the east of the country, particularly offshore, prove, as some claim, to be as large as 150 billion barrels. At present it is impossible to say whether these claims of large reserves are justified. If they are, the pressures working against conservation will be intensified, for to the pressures arising from the lot of the poor will be added those arising from the requirements of the USA, particularly should a crisis occur in the Arabian Gulf. In the latter eventuality, the US might be able to force Mexico to produce 5.0 to 6.0 mbd by the late 1980s. It is the message of this chapter that a main determinant of the rate of depletion is social pressure upon the government, whether dictatorial or elected: the faster the rate of depletion the greater the likelihood of an economic expansion which will at least maintain, and at best improve, the standards of the poor. In the light of this pressure it is thought probable that the Mexican Government, later or sooner, will abandon its conservative attitude and attempt to extract more oil than so far contemplated. In doing so it will encounter difficulties met with elsewhere in economies fast expanding their oil production: inadequate infrastructure, resulting bottlenecks and accelerating inflation. In the light of the balance a probable rate of production in Mexico in 1982 will be little over 3 mbd – scarcely enough to make good the shortfall in Iranian output.

Table 6: *Mexican production and trade in crude petroleum and refined products*

	Thousand barrels per day				
	Crude oil production	Crude oil exports	Crude oil imports	Exports of refined products	Imports of refined products
1950	204	34	0	3	10
1955	252	16	0	68	39
1960	279	3	0	21	16
1965	332	7	0	45	14
1970	432	0	0	52	32
1971	430	0	1	23	44
1972	445	0	27	15	43
1973	467	0	67	16	73
1974	595	16	18	17	51
1975	741	97	0	7	55
1976	831	98	0	3	37
1977	991	204	0	4	18
1978	1,136	302	0	8	14
1979	1,400(a)	na	na	na	na

na: not available
(a) estimate.
SOURCES: UN, *World Energy Supplies 1950–1974*, pp. 207, 306; UN, *World Energy Supplies 1973–1978*, pp. 129, 164; *Petroleum Economist*, Volume XLVII, Number 3, March 1980, p. 135.

Finally there is Ecuador, a newcomer, like the United Kingdom, to the field of oil; but since the discoveries are so recent the institutions connected with oil reflect ideas current in the second half of the 1970s and are not legacies from an earlier era. The first significant discovery was in 1967 and exports began only in 1972. Ecuador became a member of

OPEC in the following year. As a result there have never been concessions to private oil companies. From the beginning the discovery, the exploitation, the transportation, and the processing of Ecuadorian oil has been in the ownership of a State company – the Corporación Estatal Petrolera Ecuatoriana (CEPE). CEPE may, however, enter into arrangements with private companies for whatever purpose is considered desirable – there is a contract, for example, with Texaco. The arrangement, however, is different from those obtaining in the Middle East, where the State has entered into possession little by little; in Ecuador the State was the master from the outset, with the personnel being almost entirely indigenous.

Even so, there have been problems with the oil companies. When exports began in 1972 the State found some of the contracts not entirely to its satisfaction. The contracts were accordingly renegotiated, and the State's share of revenue from oil is one of the highest in the world.

Ecuador's policy, however, is not so much to export, but rather to channel both the oil and the revenues arising from it into internal industrial development. A military regime, pledged to introduce democracy, could scarcely act otherwise; it had to show the populace quick results. Besides, these are early days in the estimated twenty-year life of Ecuador's oil. On both counts no heed is paid to the control of depletion; the accent is on rapid industrial growth. One instrument to this end is a State development fund (FONADE), deriving its finances in part from oil. Among the projects underway are a fertilizer plant and a petro-chemical complex, to meet not only domestic needs but also those of the Latin American market. The pattern of Ecuador's foreign trade is likely, therefore, to undergo a change, the import of oil derivatives declining and possibly exports of crude declining as the domestic demand, industrial and private, increases. In 1978 the industrial sector showed a growth of 13 per cent, one of the highest in South America.

The industrial growth of Ecuador, however, has to be seen not in the narrow context of Ecuador itself but within the wider framework of the Andean Pact. The Andean Pact

Table 7: *Ecuadorian production and trade in crude petroleum and refined products*

	Thousand barrels per day				
	Production of crude petroleum	Exports of crude petroleum	Imports of crude petroleum	Exports of petroleum products	Imports of petroleum products
1950	7.2	2.5	0	0	0.1
1960	7.5	0	4.3	0	0.4
1970	4.1	0.8	19.5	0	0.5
1971	3.7	1.0	24.3	0	0.7
1972	78.1	68.8	23.3	1.5	0.4
1973	208.8	195.1	18.0	2.0	0.5
1974	177.0	164.1	26.7	0.9	0.4
1975	160.9	145.8	23.4	0.3	1.1
1976	187.8	167.6	22.7	0.9	3.9
1977	183.4	138.2	14.1	0.9	9.7
1978	202.0	144.7	12.1	na	6.3
1979	218.4	na	na	na	na

na: not available
SOURCES: OPEC, *Annual Statistical Bulletin 1977*, pp. 15, 16, 64, 83; UN, *World Energy Supplies 1950–1974*, pp. 203, 298; UN, *World Energy Supplies 1973–1978*, pp. 128, 162; *Petroleum Economist*, Volume XLVII, Number 3, March 1980, p. 135.

stands for more than the removal of internal barriers to trade. Its real significance lies in the development of joint industrial programmes, different countries contributing in different ways to the same project. The two aspects – freedom for the internal movement of goods and a joint industrial programme – are, however, complementary. Certainly internal tariffs could be abolished without the need for a joint industrial programme; but a joint industrial programme could not succeed without internal free trade. The most important project is the motor vehicle programme already mentioned in relation to Venezuela (see page 59). Other less important projects extend to steel, electronics, petro-chemicals and pharmaceuticals. If Ecuador succeeds in its ambition to industrialise, the success will be due in part to oil, but possibly in greater part to the Andean Pact, which has the advantage of containing a large population. The Andean Pact, for all the problems inevitable in an association of states, has advanced where the European Economic Community has scarcely trodden, and with two of its members – Ecuador and Venezuela – being members of OPEC, it has much to teach its Middle East partners in that organisation. This is one important reason why, in the interest of the Middle East, South America and the Middle East remain linked together in OPEC.

The general conclusion for the main oil-producing countries in South America is that, barring unforeseen discoveries which will make possible some conservation in the ground, they are likely to deplete, or to be forced to deplete, to the full, while further discoveries would depend on large investments and, therefore, on a higher price for oil.

Latin America

Allen, Loring, *Venzuelan Economic Development: a Politico-Economic Analysis,* JAI Press, Connecticut, 1977.

Betancourt, Romulo, *Venezuela's Oil,* Allen and Unwin, London, 1978.

Dinkelspiel, John R., *Technology and Tradition: Regional and Urban Development in the Guayana.*

Griffin, Keith (editor), *Financing Development in Latin America,* Macmillan, London, 1971.

Hassan, Mostafa F., *Economic Growth and Employment Problems in Venezuela. An Analysis of an Oil Based Economy,* Praeger, New York, 1975.

Lieuwen, Edwin, *Venezuela,* Oxford University Press, 1967.

Moron, Guillermo, *A History of Venezuela,* Allen and Unwin, London, 1964.

Salazar-Carrillo, J., *Oil in the Economic Development of Venezuela,* Praeger, New York, 1976.

Tugwell, Franklin, *The Politics of Oil in Venezuela,* Stanford University Press, 1975.

CHAPTER 5

Iran: the Failed Alchemy

There are two Irans: the monarchical or Shahanshahi Iran and the republican or Khomeini Iran. The second follows the first in time, and represents the opposite in concept – at least if the concept is interpreted as the rate at which the country should be industrialised at the expense of depleting the oil in the ground. The future, both for Iran and the rest of the Islamic world, may perhaps be best understood if chronology is followed. Let us then begin with the Shahanshah or the King of Kings.

The Pahlavi dynasty, through its founder Reza Shah, and subsequently his son Muhammad Reza Shah, overthrown in 1978, stood for rapid industrialisation, even if this meant exploiting the oil reserves to the full. Even in the recession of 1975–1976, some five mbd, out of the total capacity of seven mbd, were being exported. Nor does much appear to have been practised by way of secondary recovery. 'In 1973 the crude recovery rate in southern oilfields was 15–35 per cent, depending upon the oilfield. NIOC (the National Iranian Oil Company) planned to spend $500 m. in the next three years to increase the recovery rate.'[1] But secondary recovery through the reinjection of gas or water is an expensive business, and in view of the subsequent financial strains on the country it is doubtful whether any was undertaken. Be that as it may, the policy of going all out for industrialisation was not determined by any fine economic calculation: it arose from the simple fact that successful industrialisation spelt power. And power in its turn meant freedom – first from

---

1 Fereidun Fesharaki, *Development of the Iranian Oil Industry*, Praeger, New York, 1976, p. 61. Other sources have told the author that the recovery rate did not exceed 25 per cent.

intruding Russians to the north, and second from the British in the south, where their pressure to protect the passage to India had historically been felt, and was still believed in. (There is a Persian saying to the effect that there is an English hand behind everything, and even the Khomeini revolution is ascribed by some to the machiavellian English.) Above all, after the withdrawal of the British from East of Suez, power meant command of the Hormuz straits at the neck of the Persian or Arabian Gulf, and thus the outlet of Gulf oil to the rest of the world.

It was under Reza Shah that industrialisation began to make progress, particularly from the 1930s. Only in the late 1930s did private entrepreneurs come forward, having presumably been prevented from emerging earlier by lack of surpluses in agriculture and merchanting. In their absence, the State, in the guise of the monarch, had to press forward with industrial development, establishing in the decade between 1931 and 1941 a number of factories concerned almost exclusively with the processing of domestically owned, or mined, products. As expressed many years later: '. . . our experience here in Iran has convinced us that economic recovery and economic development have come only in the wake of government initiative and enterprise. Without wishing to generalize, and without any desire to minimize the significant role of private enterprise in the national effort for development, it seems to me that in the emerging societies and in societies that have only just begun to throw off shackles of long centuries of immobility, where the tradition of private enterprise is, to say the least, weak, by far the greater burden falls on the public authority. Not only does it have to plan; not only does it have to look after the infrastructure; it must also take the lead in industry, at least in some basic industries, and in breaking new grounds.'[1]

Under Reza Shah there was no formal plan; this emerged only after the Second World War under his son, Muhammad Reza Shah. Although indecisive, Muhammad Reza Shah was

1 Mr Mehdi Samii, former Governor of the Central Bank of Iran, quoted in *Iran in the 1970s*, Iran Chamber of Commerce, Industries and Mines, Tehran, 1971, p. 106.

ambitious. The Plan Organisation which he created started effectively in 1946, though the First Plan (1949–56) was not published until 1949. An interesting witness and assistant to the New Organisation was a 'consortium of American engineering firms, originally formed to work in Japan . . . and hired to conduct a thorough survey of Iranian investment requirements and the adequacy with which the proposed Seven Year Plan would meet them.'[1] The group delivered its report after the passage into law of the First Plan and in this sense its work was abortive; its existence could, however, come to be endowed with a wider significance, a means of drawing up for a developing country a coherent industrial programme.

Between 1949 and 1978, in roughly thirty years, the Plan Organisation gave birth to five plans, the first two of seven years each, the last three of five years each. The increase in oil prices in 1973–74 took place in the middle of the Fifth Plan; as a result new projects were added to the Plan. The Sixth Plan was to have been published on Nowruz, or New Year's Day, March 21, 1978. The Shah was then still in power, but the Plan did not appear. It has still to see the light of day. It is indeed doubtful whether in the last two years of his reign the Shah was effectively governing – subsequent events make it seem possible that the absence of effective government may have been due to illness. It cannot be said that each plan attained its aims, but even so, the gross national product had increased at a phenomenal pace since planning began. Yet clearly something had gone wrong; The explanation lies not in the nature of planning, but at the very heart of the problem of transforming a non-developed into a developed society through oil.

To begin with, there was an ambiguity in the role of the Plan Organisation itself. Was it meant only to draw up the broad framework of the Plan, and fit into it projects, suitably trimmed, worked out by individual Ministries or regions? Or was it meant to draft the projects which were to be the

---

1 George B. Baldwin, *Planning and Development in Iran*, Johns Hopkins, Baltimore, 1967, p. 31.

components of its macro-plan? Theory required it to limit itself to the first role – outlining the framework. Practice in a State accustomed to centralised power compelled it also to take on the second role – the writing of individual plans. Inevitable though this was for historical reasons, the assumption by the Plan Organisation of this second role stultified in Ministries and regions the capacity to evolve plans of their own. Throughout the existence of the Plan Organisation the unresolved tussle continued. Towards the end of the Shah's regime a renewed attempt was made to shift power down to Ministries and regions. The attempt was, however, short-lived; the Ayatollah Khomeini supervened, and whether he has the power to turn the course of history time alone can tell.

The second problem facing the Plan Organisation was money. Where was it to secure the wherewithal for its investment projects? The answer was on the face of it easy – oil. The first concession had been granted in 1901 to an Australian named d'Arcy, who, seven years later, did indeed strike oil, and sold its exploitation to the Anglo-Persian Company, later the Anglo-Iranian Company. If the Persian Government were to industrialise the country, it had to lay its hands on some of that money. The first step was taken in the early 1930s, as a result of the world-wide depression and the accompanying decline in the share of profits accruing to the Iranian Government: the area of the concession was reduced and the monies exacted for the Government were in various ways increased.

The next step was taken in 1951, when the then Prime Minister, Dr Mossadegh of the National Front (a party which has latterly jumped into renewed prominence) nationalised the company. To the outside world, Mossadegh was a strange, pyjama-clad figure, but to his countrymen he was a 'near-charismatic national leader who for all his faults effectively challenged an unfair economic institution vital to the nation's health, who stood for full implementation of the constitution, for honesty in government, and for a much

Iran: the Failed Alchemy

greater measure of intellectual and political freedom than Iran has ever known, before or since.'[1]

But why should the institution known as the Anglo-Iranian Company be regarded as so unfair? Because 'There is no evidence to suggest that [it] was involved in any local [manufacturing] venture in Iran . . . The company had simply no other interest in Iran, except for those in its concession agreement'[2] – exactly the same argument as was encountered in the case of Venezuela.

Not that the act of nationalisation spelt at that time a drop in world oil supplies. The major international firms (the so-called 'seven sisters') still retained sufficient command over the world-wide oil scene to make good the fall in output. Nationalisation eventually meant the vesting of all assets in the National Iranian Oil Company, with extraction, refining, etc being undertaken by a consortium of firms. The Anglo-Iranian Company, rebaptised BP (which stood both for Benzina Pars and British Petroleum) had a share of 40 per cent in place of its previous 100 per cent. Each company was required to pay to the Government 50 per cent of its profits, which included a royalty related to the price of crude oil.

The immediate consequence of nationalisation was a reduction in the amount of oil taken by the international oil companies, as a retaliatory measure against Iran. Table 8 shows how production fell dramatically in 1951 and subsequently, until 1954. This latter date, the year after Mossadegh had been overthrown, is important, for a new agreement between the consortium and the Iranian Government formally confirmed the act of nationalisation but left effective control of prices and output in the hands of the international oil companies. From 1955 production increased steadily, but Iran, like other OPEC members, had to struggle through the 1960s to secure an increasing share of oil revenues.

A further step was taken in 1957, when the remainder of

1 Baldwin, op. cit., p. 19.
2 Fesharaki, op. cit., p. 20

the country was opened up for exploration. One of those to take advantage of the enlargement was the buccaneer Italian, Mattei, who agreed to a division of profits of the order of 75 per cent to the Iranian Government and 25 per cent to ENI, the Italian State oil company. He was capped nearly ten years later by the French State company, ELF-ERAP, which agreed that up to 91.5 per cent of all profits should go to Iran. OPEC, with all its power, had by then come into being, the Shah being one of the most hawkish of its members. Indeed, Iran has always been one of the most prominent and forceful members of OPEC. As a country with a relatively large population in need of oil revenues Iran has taken the lead in seeking to raise prices. The Shah presided over the 1971 OPEC Tehran meeting, which effectively wrested the control of prices from the oil companies, and placed it in the hands of governments. There is evidence that this success went to the Shah's head. In 1976, Iran split OPEC by raising prices by 10 per cent, while Saudi Arabia raised prices by only 5 per cent. The Khomeini regime in its turn has taken an aggressive stance within OPEC; on oil prices, at least, it has merely continued the Shah's policy.

The Plan Organisation, then, did not lack monies for the execution of its projects, which were indeed bold. Nor was its personnel lacking in talent. But how the monies were administered depended on the Ministry of Finance, and 'the traditional attitude of the Ministry of Finance has been that its job ends with raising money; it is up to "the government" to decide how to spend it'.[1] The Ministry, therefore, has always been a weak controller of government spending (and has remained unreformed to this day). Now, a development or investment plan carries with it increases in current expenditure – for example, in education, health, road maintenance. The Ministry of Finance proved itself ineffective in controlling these. 'The revised [i.e. increased] Fifth Plan anticipated a ratio of current to capital costs of about 1 to 3 in the non-defense categories. But in 1974–75 they were in fact

---

1 Baldwin, op. cit., p. 45.

Table 8: *Iranian production and exports of crude petroleum and refined products*

	Million barrels per day		
	Crude oil production	Exports of crude oil	Exports of refined products
1950	0.7	0.2	0.5
1951	0.3	0.1	0.2
1952	0.03	0.00	0.00
1953	0.03	0.00	0.00
1954	0.06	0.04	0.03
1955	0.3	0.2	0.1
1960	1.1	0.7	0.2
1965	1.9	1.5	0.2
1970	3.8	3.3	0.2
1971	4.5	4.0	0.2
1972	5.0	4.5	0.2
1973	5.9	5.3	0.1
1974	6.0	5.4	0.2
1975	5.4	4.7	0.2
1976	5.9	5.2	0.1
1977	5.7	4.9	0.2
1978	5.2	4.5	0.2
1979	3.1	na	na

na: not available.
SOURCES: OPEC, *Annual Statistical Bulletin*, 1977. pp. 20, 21, 67, 77; United Nations, *World Energy Supplies 1950–74*, pp. 210, 310; United Nations, *World Energy Supplies 1973–1978*, pp. 130, 165; *Petroleum Economist*, Volume XLVII, Number 3, March 1980, p. 135.

almost 1 to 1 (0.97) and by 1975–76 had been lowered only to 0.9.'[1] In short, one of the reasons for the failure to transform through oil a traditional society into an industrialised country was the uncontrolled diversion of some of the oil monies to current spending. 'One point that seems beyond question is the need to have some kind of special institution, other than

---

[1] Theodore H. Moran, *Oil Prices and the Future of OPEC*, Resources for the Future, Washington, 1978.

the Ministry of Finance, to act as a trustee for the oil revenues.'[1]

Finally, after nearly thirty years of planning Iran was nowhere near its objective of obtaining income from exports other than the exports of oil. In the halcyon days of 1973–74 it expected to be able to continue to export oil up to capacity (7 mbd) until 1984, when there would set in a steady and irreversible decline. In the event, as a result of the world recession, exports in 1975–76 fell to around 5 mbd. Table 8 shows how production and exports fell from their peak in 1974. The further reduced level of production in 1978 and especially 1979 followed the fall of the Shah. Had the Shah remained in power it is likely that production and exports would have continued at 1976 or 1977 levels, or perhaps a little higher. However, by the mid or late 1980s the limited nature of Iran's oil reserves would inevitably have produced a decline in production.

Imports, on the other hand, continued to increase under the Shah, partly because of the capital goods required by every developing country, partly because of the rising affluence and changing tastes of the people. The migration of people from the countryside to towns changes their tastes. Food bulked large in the imports, not because agriculture was deliberately neglected, but because the attractiveness of urban living steadily denuded the villages of their manpower. And the effective organisation of agriculture has eluded Iran as well as many other countries. In 1976 only 18.5 per cent of the imports were covered by non-oil exports.

There were two reasons for the failure to develop non-oil exports. The first was that there was no measure of the productivity of capital and therefore no indication of the most propitious areas in which to invest. Concentration was entirely on the productivity of labour: The areas in which this was rising fastest, 'apart from chemical and petroleum products, [were] apparel, furniture and printing.'[2] Outside oil,

---

1 Baldwin, op. cit., p. 199
2 Aubrey Jones, *Measurement of Productivity in Iran*, Plan and Budget Organisation, Tehran, 1978, p. 28.

there were scarcely any potential exporting industries; the industries mentioned above were all labour-intensive; and labour was particularly expensive, since it was an ambition of Iran's to establish a 'welfare state', entailing high labour costs, as well as to build an industrial structure, whereas Western countries were able to industrialise with low-priced labour.

The second reason for the failure to develop sources of export income other than oil was the lack of thought given to the achievement of a technically balanced industrial whole. There was, for example, no engineering industry – only the assembling of components engineered abroad. Attention was dedicated to the forward integration from oil into oil derivatives. 'Backward linkages, with regard to capital expenditure, have been particularly weak. The oil industry is a vast capital-intensive industry, and the degree of capital intensity is rising over time by the installation of sophisticated and automatic machinery in refineries and oil terminals. The national economy has so far been unable to provide such capital equipment for the industry . . . The only major part played by the domestic economy in supplying the needs of the oil industry for capital goods has been the construction of the Ahwaz pipe mill in 1968.'[1]

Despite all the effort and expenditures (for example, imports rose from 611 million dollars in 1959/60 to 24,700 million dollars in 1977/78), Iran before the revolution of 1978/79 had not been successful in building a balanced industrial base. Industry had developed in an essentially haphazard fashion, the main aim being to substitute domestic for imported manufactures. Even this limited objective was hampered because 'the balance of resources between agriculture, construction and industry is being upset by the rapid increase in construction output and its consequential increased demands for both finance and manpower.'[2] The

---

1 Fesharaki, op. cit., p. 143.
2 Aubrey Jones, *Final Report of a Consultancy Mission on Productivity, Incomes and Prices (1 December 1976 – 30 June 1977)*, Plan and Budget Organisation, Tehran, 1977, p.4.

absorption of labour into construction rather than manufacture is brought out in Chart 3, the construction in question being almost entirely in the private sector, only minor sums of money being devoted to public housing. In Tehran, the poor lived in hovels, while the rich entertained on plates of gold. It can be personally testified that this differentiation was of great concern to the now executed Mr Hoveyda, when he was still Prime Minister. Chart 3 shows that the value and volume of construction starts increased far more rapidly during the 1970s than did industrial output. It was recognized that a country undergoing economic development should spend substantially on the construction of roads, railways, ports, water and electrical power systems, and also that the export of manufactured goods ought eventually to replace revenue from oil; yet Iran's industry remained concentrated on assembly, and on areas which had a low value-added content and little potential for export.

It has been said, particularly in other Islamic countries, that the Shah tried to develop too fast. It is true that he was seeking to develop at a pace faster than that to which the social habits of the country could adapt themselves. But in addition he was developing in the wrong way, neglecting the basic engineering industry without which there cannot be a modern industrial society. A full industrialisation would have required the co-ordination of investment, tariff and pricing policy, and there was no such co-ordination. Nor was there, despite the welfare state, a serious attempt to abate the disparity in incomes between rich and poor. In 1970 the disparity in incomes between the upper and the lowest tenths was 19 to 1, an indication that the wealthiest were spending eighteen times more than the poorest. By 1975, the year of fastest economic growth, this ratio had increased to 38 to 1. Similarly the growth in expenditure between the same two years shows that the wealthy had increased their spending by 3.6 times and the poorest by only 1.7 times. In other words, the poorest just kept pace with the rise in the cost of living, while the rich increased their real spending (that is, their

Iran: the Failed Alchemy

Chart 3: Indices of construction starts, and construction wages, compared with industrial output

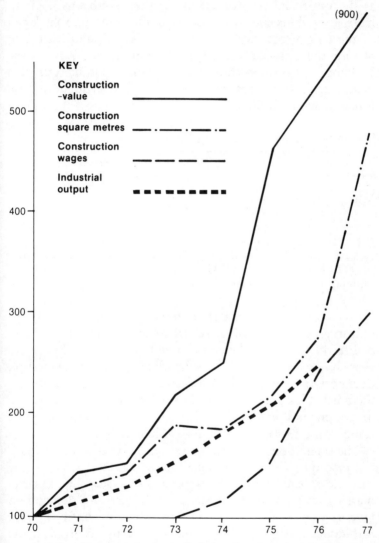

SOURCE: Aubrey Jones, op. cit., Appendix 5; from Iran Central Bank.

spending in terms of constant prices) by at least a factor of two.[1] The story is the same as that of Mexico.

Suppose the Shah had survived; would he, from his palace at the upper skirting of the smog that overhangs Tehran, have realised his cherished ambition of making Iran the Japan of the end of the twentieth century? It is doubtful, for, in chasing the glamour of derivatives from oil, not only was he building a lop-sided industrial structure, but he was also venturing into an area of likely decline, as is suggested by Table 9, which follows:

Table 9: *Growth rates of major petrochemical blocks*

	Growth rates per annum; per cent
1950–1960	15.5
1960–1970	17.0
1970–1980	8.5
1980–1985	6.0
1985–1990	5.0

SOURCE: Lecture given by Mr L. R. Burchell of BP Chemicals Limited, to OPEC Seminar, October 5, 1977.

Does a country which has attempted to transmute oil into industry, and then failed, revert to the condition it was in before? No, because the clock cannot be put back. For thirty years Iran experienced increasing affluence and a proliferation of expectations which, once released, cannot be put back into the bottle. Such expectations live on frustrated, muttering in discontent and finally striking out in unrest. In Iran the unrest came in the form of the Ayatollah Khomeini.

There had been plenty of warning. In the autumn of 1977, for example, from one's penthouse flat overlooking the garden of the Water Board one could see the workers massing in front of the managerial offices and demanding higher wages, though legally such demonstrations were forbidden. Early in 1978 a similar incident occurred at the assembly plant of General Motors. On that occasion the

1 Aubrey Jones, op. cit., pp. 34, 35.

Iran: the Failed Alchemy

gendarmes were there, with guns at the ready. On the arrest of the apparent ringleader, the massed workers moved threateningly nearer the offices. The Iranian General who was supposed to act as an intermediary between General Motors and the Iranians had not been seen for a week, so frightened was he of the confrontation to come.

A revolt against the regime could thus have been predicted, even if its exact nature might not have been discernible. The omens lay in the gross disparity in incomes, in the rate of inflation (measured by the author at 40 per cent a year, a figure not refuted), in the ostentation with which wealth was displayed – and above all in the nature of Islam itself.

Islam cannot be said to be against innovation – indeed there is probably more written about change and innovation in the Quran than in the New Testament. Islam is certainly against usury – that is, an extortionate rate of interest – but it is not against profit in the sense of a reward for risk taking. Nor has Islam become institutionalised as much as the Christian Church. Nonetheless, as Islam has developed it has acquired the aspect of resistance to change. Scarcely anywhere is this better expressed than in the following quotation: '. . . the most important single obstacle to progress, in the world of values and attitudes, remains the deep-rooted feeling that what is old is good, that change is bad, and that progress, or, to be more precise, improvement, consists in restoring what existed before the change . . . A good example from Turkey may be found in the analysis and recommendations of the long series of writers and observers who examined the causes of the decline of Ottoman power and propounded remedies . . . all of them, without exception, saw the basic fault as a falling-away from the high standards and good practices of the Islamic and Ottoman past, and the basic remedy as a restoration of those standards and practices.'[1]

It is to be recognised that later many Muslims came to see this return to the past, this recovery of an earlier state of perfection, as a mirage. But is this true of the Ayatollah

---

1 Bernard Lewis, *Islam in History*, Alcove, London, 1973, p. 301.

Khomeini? It is doubtful; and if his regime continues to cling to the mirage, what does that imply in terms of oil?

Initially, in terms of production, the statement of a decline in Iranian oil output in the autumn of 1978 to between the claimed output of 3 mbd to 3.5 mbd seems, after an examination field by field, to be reasonably valid. The political-religious atmosphere, meaning in economic terms a policy of nil growth, makes it most improbable that production will again be raised above this level. The crucial question is whether it can be maintained. By early 1980 the services of the consortium, which was supposed to help with the production and marketing of the oil belonging to the Iranian Government had been dispensed with. Whether on account of this or because of the natural deterioration of the reservoirs production subsequently fell,. By the summer of 1980 it was around 1.7 mbd. The most one can say is that, if an output of 1.5 mbd to 2.5 mbd is maintained over the next two years, there is no reason why it should not be maintained for some years longer. If, however, there is a further decline between 1980 and 1984, it is probable that Iranian oil supplies will quickly dwindle to nothing. Such a reduction would be the result of inability to service and maintain production and export facilities.

The likelihood of a further decline in Iranian oil production is reinforced by the attitude of the Khomeini revolution towards administrative authority. Each Ministry has its purge committee of some five members, some of them clerks with no more than two years' experience. All those with fifteen or more years' experience are made redundant, as also are many talented officials who have served for less than fifteen years. It is estimated that the staff of all Ministries will soon be halved. Authority has been bestowed on the lowest level. This is not participation in decision-taking by the bottom, but the wielding of absolute control by the bottom. The present disorder is an inevitable consequence and has reached a point where even the rich, their riches now gone, would see salvation in an orderly Communism. Against this background the omens for oil production are singularly unpropitious.

Iran: the Failed Alchemy 81

What of the effects of the Ayatollah Khomeini on the rest of the Islamic world, and in particular on its oil-producing members? At a superficial view it might be expected to be slight because Khomeini is a Shi-ite (one who believes that Ali, the Prophet's son-in-law, should have been his successor) while the rulers of the rest of the Islamic world are Sunnis (those who accept the Sunnat, or history, as it actually occurred), and why should they accept the influence of the leader of a minority sect? But Islam is Islam, whether Shi-ite or Sunni; and an Islamic event as important as Khomeini's revolution can hardly fail to exert an influence on the rest of the Islamic world, even if that influence may not be immediately evident.

The most important effect is likely to be on the volume of oil output. Rightly or wrongly, there is a feeling that under the Shah Iran exploited its oil resources too fast. And attempts to develop fast are likely to fail in all Islamic countries. Islamic tradition (though not necessarily the Quran) lays emphasis on the word as against the work of the hand – witness the respect paid to poets as distinct from graphic artists – so the kind of manipulative skills necessary in industry have to be developed, and this takes time. Furthermore, Islam's great scientific past has been obliterated by the passage of time. These would be good reasons for other Islamic countries to follow Iran's example and slow down the development of their oil resources.

More important, however, than the effect of the Khomeini revolution on the future rate of oil production, is its possible effect on the politics of the Islamic world. For the Khomeini revolution was unique in the world's history, in that it combined both populism and religion. In its populism it reverses the traditional hierarchy in which power flows from the top. And in its religion it has shown a cruelty which is alien to all Islam, Shi-ite and Sunni. Perhaps it seeks to return to the state of perfection which Christians see in Heaven, but which Muslims see 'in the past, in terms of mythology, a revelation or master-philosophy. Given this original perfection, all change is deterioration – a falling-away from the sanctified past.'[1] But who is responsible for the deterioration? The Ayatollah Khomeini might well reply: 'secularist regimes'. If other Islamic countries come to give expression to this outlook, the political implications can be profound, there can be no development and, unless appetites have been sufficiently aroused, little oil for the materialistic Western world. In the chapters that follow an attempt will be made to detect such incipient signs as there may be of a change in view, though the signs are likely to remain confused.

Iran

Baldwin, George B., *Planning and Development in Iran*, Johns Hopkins, Baltimore, 1967.

Fesharaki, Fereidun, *Development of the Iranian Oil Industry*, Praeger, New York, 1976.

Iran Chamber of Commerce Industries and Mines, *Iran in the 1970s*, Tehran, 1971.

Jones, Aubrey, *Final Report of a Consultancy Mission on Productivity, Incomes and Prices (1 December 1976 – 30 June 1977)*, Plan and Budget Organisation, Tehran, 1977.

Jones, Aubrey, *Measurement of Productivity in Iran*, Plan and Budget Organisation, Tehran, 1978.

Lewis, Bernard, *Islam in History*, Alcove, London, 1973.

Moran, Theodore H., *Oil Prices and the Future of OPEC*, Resources for the Future, Washington, 1978.

1 Bernard Lewis, op. cit., p. 294.

CHAPTER 6

Saudi Arabia: the Conflict within Islam

Saudi Arabia, so named after the dynasty which has ruled her for two centuries, Al Sa'ud, has the largest oil reserves in the world, the largest on-shore field in the world (Shawra), the largest off-shore field, and after the Soviet Union is the largest producer in the world. Her fewer than 800 wells, all in the east, around the Arabian Gulf, have a daily output of 12,000 barrels each, as contrasted with the United States, where roughly 500,000 wells each produce on average only 16 barrels a day. 'We are a very appealing piece of cake. There are lots of people who would dearly love to get a slice of us.'[1]

Yet Saudi Arabia finds herself faced with some difficult problems. Is she, just because of her sheer wealth in oil, to produce more than she needs for herself to help keep the world economy moving, possibly to her own ultimate detriment? Can she maintain her position as the spiritual and geographical heart of Islam, containing as she does both Mecca where the Prophet Muhammad received his inspiration and Medina where he sought retreat on being mocked, when passing through 'a psychological and moral crisis, produced by a hysterical wave of materialism, where wealth and not morality is the standard of honour and respect'[2]? Can she keep her traditional purity when under invasion from some millions of foreign workers? Can she integrate an

---
1 Muhammad Abdu Yamani, Minister of Information, quoted in Adeed Dawisha, *Saudi Arabia's Search for Security*, Adelphi Paper No. 158, International Institute for Strategic Studies, London 1979, p. 7.
2 S. S. Husain and S. A. Ashraf, *Crisis in Muslim Education,* Hodder and Stoughton, Jeddah, 1979, p. 20.

educational system in which the schools aim at the inculcation of piety and of revealed knowledge, while the universities, based on an a-religious Western model, teach only knowledge acquired by man? Finally, can she retain her friendship towards the United States, to which she looks for her defence, while expressing her disapproval over American betrayal of the Palestinian Arabs, whose spirtual head she considers herself to be? These social, political and religious issues all affect the rate at which Saudi Arabia depletes her oil and the rate at which she can industrialise.

Table 10 shows how Saudi crude oil production and exports have risen steadily since 1950, to a level of 9.2 and 8.6 mbd, respectively, in 1977. A lower level of world economic activity in 1978 was reflected in a drop in Saudi production to 8.3 mbd. In 1979, to the accompaniment of much internal criticism, the rate of production was raised to 9.5 mbd, a rate continued through 1980. At this level, production was almost up to the limit – 12 mbd – of 'sustainable' capacity: that is capacity which can be kept constant given existing facilities. Unless further discoveries of oil are made, Saudi Arabia could be on the eve of a decline in the ratio of reserves to production. Superficially, the increase in production in 1979 could be justified on the ground that in 1977 reserves rose from, roughly, 151 billion barrels to 169 billion barrels; since then they have dropped slightly – to 167 billion barrels in 1978 and 168 billion barrels in 1979. If the present level of production is maintained – and it would be difficult against an adverse world opinion to decrease it – and no further large discoveries are made, Saudi's oil reserves could last some fifty to sixty years, scarcely long enough to change the social habits of a traditional society which began to industrialise only around the early 1970s.

Why did the Saudi authorities decide to run so risky a course? Subsidiary reasons could have been the decline, deliberate and not accidental, in Iranian production and, possibly, pressure from the USA. Certainly such pressures had been exerted before. For example, Mr James Akins, US Ambassador in Jeddah, is reported to have said in 1974: 'The

Table 10: Saudi Arabian production and exports of crude petroleum and refined products

	Million barrels per day		
	Crude oil production	Exports of crude oil	Exports of refined products
1950	0.5	0.4	0.1
1955	1.0	0.8	0.2
1960	1.3	1.1	0.2
1965	2.2	1.9	0.2
1970	3.8	3.2	0.4
1971	4.8	4.2	0.4
1972	6.0	5.4	0.3
1973	7.6	7.0	0.3
1974	8.5	7.9	0.3
1975	7.1	6.6	0.3
1976	8.6	8.0	0.4
1977	9.2	8.6	0.5
1978	8.3	7.7	0.5
1979	9.5	na	na

na: not available.
SOURCES: OPEC, *Annual Statistical Bulletin 1977*, pp. 27, 79; OAPEC, *Sixth Annual Statistical Report 1977–1978*, pp. 11, 15, 56; OPEC, *Annual Report 1978*, p. xiii; UN, *World Energy Supplies 1970-1974*, pp. 212, 312; *Petroleum Economist*, Volume XLVII, Number 3, March 1980, p. 135.

purpose is to show the Saudis that it is worth producing oil rather than keeping it in the ground.'[1] Mr Kissinger, for his part, is reported to have said in December 1973, while he was still Secretary of State, 'a massive effort should be made to provide an incentive for the producers to increase their supplies'[2]. In spite of these unwise urgings, which, if listened to, would have destroyed any hope of development for the oil-producing countries, it does genuinely appear to be the case that the Saudis are conscious of their membership of a world community and would not regard it as being in their

---
1 Michael Field, *A Hundred Million Dollars a Day*, Praeger, New York, 1976, p. 73.
2 Ibid.

own interest to jeopardise that community's economy. Their leader in this respect is Sheikh Yamani, the Minister for Petroleum and Minerals. He is one of the few Ministers not of royal blood, and the title 'Sheikh', which, strictly speaking, means tribal chief, is honorary. Soft of speech, he is far of sight, and his seemingly infinite wardrobe of robes arouses both awe and frustration in his officials, who cannot possibly keep up with him.

The domestic risk which Saudi Arabia is running in producing and exporting so much is all the greater when the rate of increase in internal consumption is considered. In the late 1970s this was running at a rate well over 20 per cent, and there were moments in the mid-1970s when it touched 30 per cent. While a high rate of consumption is only to be expected in a developing country, particularly one without railways and with little public transport, it is also due in part to the low domestic price for oil and its immediate derivatives charged to both private and industrial users.

This low price, too, is a source of internal controversy. Those who are opposed to it contend that it cannot go on for ever; nor indeed can it if the ratio of reserves to production is to enter into decline, as at some stage it must. In addition to the low price of oil, food is subsidised, and the two in combination, it is contended, make people work-shy. Those, on the other hand, who are in favour, see the low price as a means of palliating the differences between rich and poor, of mitigating the increase in the cost of living, which trebled between 1970 and 1977, and of nurturing infant industry. In addition, they see it as a proper use of Government revenues which depend almost entirely on oil – oil company payments amounting in 1979 to almost 50 billion US dollars, while revenues not stemming directly or indirectly from oil are negligible. It is evident that the controversy over the internal price of oil and its immediate derivatives goes to the heart of Saudi's social problem.

The oil companies are three: the Getty Company, operating privately in the neutral (that is, disputed) zone between Saudi Arabia and Kuwait; the Arabian Oil Company, a 50–50 arrangement between the Saudi Government and the Japan Petroleum Trading Company, operating in the same area; and, dwarfing the first two, Aramco (or the Arabian American Oil Company), which operates in and off the Arabian Gulf, and has grown by accretion to comprise four partners, Texaco, Socal (or Standard Oil of California), Exxon and Mobil. In addition there is Petromin, the Saudi National Oil Company, created in 1962 with power to develop the petroleum, petrochemical and mineral industries of the nation. It is this triangle of the Saudi Arabian Government (represented by the Ministry of Petroleum and Mineral Resources), Petromin, in effect the Government's arm, and Aramco which is intriguing.

The relationships between the Saudi Government and Aramco have undergone roughly the same evolution as has occurred in other oil-producing countries between the relevant governments and the oil companies, though with less acrimony than elsewhere – a 50–50 profit sharing arrangement in 1950; an agreement in 1951 that the new formula should apply before the payment of US taxes rather than after, to the benefit of both the Saudi Government and the consortium (to the Government in that it had a larger share of the profits and to the consortium in that, to the US Inland Revenue, the share of profits payable to the Saudi Government was a cost deductible before the computation of US tax); the imposition by the Saudi Government around roughly the same time of a 50 per cent tax on net profits (that is, profits after they had been shared with the Government) in addition to the payment of a royalty at the rate of 22 cents a barrel; the introduction in 1959 of two Saudi Government representatives to the Aramco board; and the agreement in 1965 that royalties should be allowed as an expense before calculating the 50 per cent income tax, rather than being deducted from the tax due.

The Saudi Government had appeared to secure a victory in 1955 when Aramco agreed to base taxes on posted prices, not the realised prices at which it sold oil to the parent companies at a discount. However, the companies reduced the posted prices throughout the 1950s, and thereby restricted the Government's tax revenue. Thus, the next step forward did not come until 1971, after the Tehran Agreement, with the passage to the Government from the consortium of the right to declare posted or published prices. Then came the beginning in 1972 of Government participation in the equity of Aramco, at a level of 25 per cent, culminating in fact, if not in form, in complete ownership by the Government on January 1, 1976, a date earlier than that previously agreed upon. The speed with which complete Government ownership was effected may perhaps explain the fact that there was no reduction in investment by the oil consortium as had happened elsewhere.

The difference between the form and the fact is that, by the spring of 1980, the treaty governing complete public ownership had still not been signed – that was the form; but whatever the date of signature total Government ownership was to be retrospective to January 1, 1976; from that date, therefore, the Government was the complete owner – that was the fact. From that date also Aramco, from being a concessionaire, became a consortium operating the oil and gas fields on behalf of the Government.

The as yet unsigned treaty is believed to contain the following main provisions:

1. Aramco is paid a fee for its part as an operator.
2. In addition, it receives a bonus or 'entitlement' if it discovers fresh reserves, the 'entitlement' to each consortium member being proportionate to the original shareholding. This provision is believed to be unique and, though the agreement is still unsigned, the Saudi Government has consistently honoured the obligation. If the discovery of new oil is effected by Petromin, Aramco receives no 'entitlement'.
3. The members of the Aramco consortium have first claim on the crude oil they produce for their own refineries and petro-chemical plants. Other private buyers take second place.

4. Petromin is the sole seller within Saudi Arabia of Aramco-produced oil, as well as the sole seller to foreign governments.

Relationships between the Saudi Government and Aramco do not end with the signature of a formal treaty of nationalisation. In spite of all the wranglings of the lawyers, the completion of the treaty is probably the easiest step. Beyond the treaty the Saudi Government is interested in two vital issues: the management of the reservoirs in such a way that there is no waste; and the training by the American operators of Saudis to take their place, for the act of nationalisation is not consummated until the operation is taken over by nationals. Petromin plays an important part in both issues. As far as the management of the reservoirs is concerned, it has representatives in the field, monitoring each activity; the Ministry of Energy also has its own observer. And as far as the training of top managers, rather than just of operatives, is concerned, Aramco started a training programme after pressure by Petromin lasting two years. Again the efforts of Petromin are reinforced by the Ministry of Energy.

It is a theme of this book that to an oil-producing country oil represents the equivalent of agriculture to industrialised countries – it is the source of the surplus without which there can be no industrialisation. Table 11 shows that in recent years – years during which Saudi Arabia has made a vigorous attempt at industrialisation – oil revenues have contributed over 85 per cent of all government revenue. It is the scale of oil revenues since the price rises of 1973–74 which has provided the spur to and means for Saudi industrialisation. Further, the faster that industrialisation can proceed, the more slowly need the oil be depleted as an alternative income would become available; conversely, the faster the rate of depletion, the more likely is it that industrialisation will be retarded. What judgement can be made about the relationship between the pace of industrialisation in Saudi Arabia and the rate of depletion?

An answer is not easy, for the second five-year plan

(1975–80) was terminated in the early months of 1980 and no evaluation is likely for a considerable time. However, an official evaluation was made of the first two years of the plan – up to 1977. At that stage, the evaluation had to be confined to the non-oil manufacturing sector, industry based on hydrocarbons having yet to be developed. According to this evaluation, in 1975 the contribution of manufacturing industry to the Gross Domestic Product was less than 1 per cent; over the ensuing two years the number of privately owned factories increased substantially, thanks to generous government aid (cheap land, loans at nominal rates of interest, free imports and a tax holiday for ten years). Correspondingly, the share of oil in the Gross Domestic Product fell from 82 per cent in 1975 to abut 70 per cent in 1980.

Table 11: *Saudi Arabian Government budget revenues and expenditures*

	Actual revenues and expenditure in million Saudi riyals			
	1975/76	1976/77	1977/78	1978/79
Revenues	103,384	135,957	130,659	132,871
Oil	93,481	121,191	114,042	115,518
Other	9,903	14,766	16,617	17,353
Expenditures	81,784	128,273	138,027	147,400

SOURCE: *Middle East Economic Survey*, Volume XXIII, No. 25, 7 April 1980, p. II.

Alternative data on the progress of the Saudi economy can be seen in Table 12, also derived from an official source. The contribution to GDP made by mining and quarrying, which in this context means oil, may appear rather low, but a substantial proportion of construction and service activity is related to and dependent on this sector. The contribution of non-oil manufacturing remained below 2.6 per cent in 1977/78, although value added in this sector had more than trebled in real terms since 1968/69. The Saudi economy remains dominated by oil and services, and has been unable to diversify into manufacturing, despite billions of dollars spent on the private sector by the Government in the 1970s through such bodies as the Saudi Industrial Development Fund. Yet Table 12 shows that growth in real GDP has nonetheless been rapid, so why should manufacturing still be relatively insignificant?

Table 12: *Saudi Arabian Gross Domestic Product by economic activity*

	GDP in producers' values at constant 1969/70 prices					
	Million Saudi riyals (a)			Percentage share of GDP		
	1968/69	1976/77	1977/78	1968/69	1976/77	1977/78
Agriculture, forestry, fishing	957	1,282	1,359	6.0	3.2	3.2
Mining and quarrying	7,135	19,986	19,796	44.9	50.4	47.2
Manufacturing:						
petroleum refining	1,016	1,523	1,591	6.4	3.8	3.8
other	385	956	1,103	2.4	2.4	2.6
Construction	1,028	4,146	4,582	6.5	10.5	10.9
Services	5,108	11,425	13,209	32.1	28.8	31.5
Total GDP	15,904	39,668	41,904	100	100	100

(a) One US dollar equals approximately 3.5 Saudi riyals.
SOURCE: *Statistical Year Book*, Fourteenth Issue 1398AH–1978AD, Kingdom of Saudi Arabia Ministry of Finance and National Economy, Central Department of Statistics, Table 10–27, p. 418

Despite rapid industrial development 'Widespread under-utilisation of installed capacity indicates the gross inefficiency of the non-oil manufacturing in the Kingdom . . . Among the various reasons . . . the shortage of manpower has been reported to be most important.'[1] The report went on to say:

1 *Industrial Structure and Development in Saudi Arabia*, Industrial Studies and Development Centre, Riyadh, 1977, p. 75.

'It is obvious that Saudi Arabia will have to remain dependent on expatriate manpower for a long time for its economic and industrial development.'¹ It is clear that there are deficiencies in manpower in all sectors of the economy. In manufacturing in the mid-1970s the average manpower shortage for all industries represented 35 per cent of total requirements.² This is despite the fact that expatriate labour continues to dominate employment in Saudi Arabia. Of total private sector employment of 348,188 in 1976, 220,845 were non-Saudi and only 127,343 Saudi.³ As elsewhere in the Middle East, this large foreign labour force is creating social and political problems. Moreover, the emphasis placed on education and training in the Development Plans will not have much impact on the supply of Saudi labour for several years. The Second Plan expects the Saudi labour force to increase from 1,286,000 in 1975 to 1,518,000 in 1980; the non-Saudi labour force is expected to rise from 314,000 to 813,000.⁴ These figures certainly underestimate the actual foreign labour force.

Industrialisation normally starts with the production of consumer goods, protected from the outside world by a tariff. 'In Saudi Arabia whatever industrialisation has taken place even in the non-oil sector, most of it has been in intermediate products such as the production of construction materials.'⁵ The reason for this particular pattern is not clear, but construction materials are scarcely likely to form the basis of a future exporting industry to take the place of oil. The industry that there is caters only for the small internal market – the census of 1974 estimated a population of some seven million, a figure which has since clearly grown. It is to be noted also that, as in Iran, there is no engineering industry; cars are assembled, but the components are imported; and an engineering industry is the backbone of an industrialised society.

1 Ibid, p. 78.
2 Ibid, Table 4.4.11
3 *Statistical Year Book,* (op. cit. in Table 3) p. 448.
4 *Saudi Arabia's Search for Security,* op. cit. p. 14.
5 *Industrial Structure and Development in Saudi Arabia,* op. cit., pp. 76-77.

The 1970s saw a greater concentration on manufacturing industry, both in the official plans and in practice. Government policy is that industry should be the domain of the private sector, although public or joint enterprises may be necessary if the project is very large or technologically complex. At a later stage, when these industries have been successfully launched, they will, according to intention, be handed over to the private sector.

The private manufacturing sector apart, there is also a State-owned section, based on cheap gas and concentrated in a State holding company with three main components: the generation of electrical power in the eastern part of the country which is entrusted to Aramco; an aluminium smelter; and an iron and steel works with a projected annual capacity of 3.5 million tons. The first component – the generation of electric power – is clearly a public utility and as such is likely to remain indefinitely in the hands of the State. As for the other two components – aluminium and steel – it is the intention to sell them to the private sector if and when they are profitable. They will certainly enjoy the benefit of cheap gas; but the ores in both cases will for some time at any rate have to be imported from abroad. It would be wise, therefore, to suspend judgement about their likely profitability.

The most promising industrial prospects are, of course, based on oil. In the early part of 1980 there were three main refineries – Ras Tamura (Aramco), Jeddah (Petromin) and Riyadh (Petromin), their main products being liquefied petroleum (that is, associated) gas, gasoline, naphtha and the middle distillates.

Total production of refined products amounted in 1979 to over three million barrels. For the future there are planned five advanced petro-chemical complexes to be based on the two industrial towns of Jubail, on the Arabian Gulf, and Yanbu, on the Red Sea. All five plants are to be owned in conjunction with Western chemical or oil companies. An agreement for the first of these plants was signed between the Saudi Arabian Government and Mobil on April 19, 1980.

The plant is to be located at Yanbu, is to use liquefied gas piped from the east, possibly at a fifth of the world price, and will produce in the first instance ethylene, with possible extensions into low-density and high-density polyethylene. It is the intention to export into the markets of the industrialised countries.

What are the prospects of success? On the debit side are probably higher costs of construction than would be encountered in industrialised countries, higher operating costs, and a start-up delayed beyond the time planned. On the credit side are the cheap cost of gas and the low cost of finance from the Government, though the further down the stream of derivatives from ethylene that one proceeds the less important does the advantage of cheap gas or feedstock become. A favourable summing-up has been expressed: 'Given that some oil producers such as the Saudis are succeeding in providing the vital prerequisite of an industrial infrastructure, and given that they appear to be attracting extremely competent companies as partners, I have every expectation that these ventures will be successful both for the host governments and for their partners.'[1] It should be added that, from the companies' point of view, part of the prospective 'profitability' of the project could be a 'bonus' in the shape of crude oil which they might otherwise have difficulty in obtaining. The ultimate result of a petro-chemical industry could thus be a form of rationing in the supply of crude oil.

A problem could conceivably arise if the projected Saudi Arabian plants added to the surplus capacity in petrochemicals, particularly in Europe, where the rate of utilisation of capacity in the late 1970s and early 1980s was below historic levels. In such an event a possible Saudi reaction might be to sell crude oil only if the European refiners agreed to operate at a still lower level of capacity, particularly since the Saudi plants would be more up-to-date than those in Europe.

On the face of it, then, there seems no reason why Saudi

1 W. C. Thomson, *A Positive Attitude to New Producers*, Shell Reprographics, London, 1979, p. 7.

Arabia should not be able, during the lifetime of her oil, to develop a successful petro-chemical industry. Insofar as she succeeds in this attempt the internal pressures, for example, the pressure of the poor, to export crude oil would be diminished, though the external pressures from, for example, the USA, would remain untouched. Nor is there any reason why a successful petro-chemical industry should not survive the oil, given that the world's petro-chemical industry uses only about six per cent of its oil. Such a survival would, however, require two conditions: first, the elimination of the less valuable uses of oil – for example, fuel oil; secondly, the development of a different feedstock for petro-chemicals. This should not be beyond the bounds of technology given that the silicon atom is not so different from the carbon atom, though whether this technology can be developed before Saudi's oil is exhausted is difficult to say.

In addition to the petro-chemical projects Petromin has plans to enlarge existing refineries and build new ones for export purposes, potentially the most promising export being fertilisers. Further there is a project for a gas gathering treatment and transmission plant which could export in liquefied form the gas which is now being flared.

Beyond petro-chemicals, there are the minerals – phosphates, iron ore and gold. The phosphates are to be found in the north and are in effect an extension of a Jordanian field, though richer in quality. The iron ore deposit is in the west, along the Red Sea: the ferrous content is high but the iron is mixed with silicate and the separation of the two is an expensive process, so an iron and steel industry in Saudi Arabia is unlikely to be economic. The gold is also to be found in the west, and the rise in the price of the metal has made economic the re-opening of mines which had been previously closed. Mineral exploitation and processing is, however, unlikely to begin before the early 1980s.

Finally, there is agriculture, a keystone of the Third Plan. There is a flourishing agriculture in the south, irrigated with water from the natural aquifer. But the supply of water, like that of oil, is finite, and the aquifer could be threatened with exhaustion. The extension, possibly even the maintenance, of agriculture depends therefore on the use of gas for desalinating sea water. And whether there is gas enough for that remains to be seen. Only three unassociated gas wells have yet been discovered by Aramco. In this respect Saudi Arabia differs from her neighbours of the Arabian Gulf, she having a preponderance of oil over gas, they a preponderance of gas over oil, and they are all separate States, with as yet little co-operation between them.

Agriculture has long dominated the lives of most Saudis, and, given enough water, it is no doubt possible to make Saudi Arabia self-sufficient in agriculture. However, Table 12 above showed how agriculture, forestry and fishing have made a reduced contribution to GDP through the 1970s. At the same time food imports have increased substantially. It may well be that it is not possible to reconcile industrial development and urbanisation with the development of an agriculture which remains labour-intensive, since workers are attracted to the towns from the land. Agricultural development, therefore, requires a large injection of capital.

Broadly speaking then, there seems no reason why Saudi Arabia should not be able to take advantage of the interlude of oil to 'develop'. The 'development', however, presupposes a minimum level of oil production to yield the requisite revenues for the Government. While these have grown steadily over the years, expenditure has grown much more rapidly. The consequence has been that after several years of substantial budget surpluses, the Saudi budget recorded a deficit of over 7 billion Saudi riyals in 1977/78, and nearly 15 billion in 1978/79. The minimum level of oil production and exports necessary to support industrialisation depends greatly on the price of oil. With production of over 9 mbd and with the large price rises of 1979, the budget will once again be in surplus in 1979/80. In addition to domestic expenditures, Saudi Arabia spends considerable sums in aid to developing countries. The effect of these expenditures, domestic and foreign, was a balance of payments deficit in 1978 of 3.3 billion riyals, the first since 1969, compared with a surplus of 43.5 billion riyals in 1977. In consequence, Saudi revenue requirements probably necessitate a minimum production of perhaps 7.0 to 8.0 mbd. To produce above this level, as at present, means that the Saudis risk acquiring temporarily surplus revenues and building up foreign financial assets which cannot maintain their value in real terms. Only in the course of time can these revenues be invested in the country.

The continued production of oil, however, and *a fortiori* industrialisation depend on more than economics. They depend on things social, political and spiritual. They depend, in short, on whether Islam, while remaining Islam, can adapt itself to the modern world.

Iran, it has been seen, failed. And there are some frightening similarities between the Iran of the Shahanshah and Saudi Arabia. There is the same frenzied development, the number of licensed factories jumping from thirty-eight in 1973 to two hundred and sixty-seven in 1976 and two hundred and fifty-eight in 1977;[1] there is the same corroding inflation, more marked in housing than in other things, the general price index for certain income groups rising between 1975 and 1976 by 32 per cent but that for housing by 44 per cent;[2] there is the growing disparity in incomes which is the concomitant of inflation, though a measurement is difficult to ascertain; and there is the same ubiquitous construction, though the boulevards are broader and the buildings, interspersed with mounds of ochre earth, are more stylish.

Yet there are also profound differences between the Iran of the Shah and Saudi Arabia. The Shah tried to secularise his country and so aroused the opposition of the main organised force, the Church. In Saudi Arabia, by contrast, Church and State are one, in accordance with the Sunna (tradition) that religious and temporal power are twins. The monarchy has codified the legal system on Quranic lines. The Council of Ministers contains two prominent Islamic theologians, as

---

1 *Industry*, Saudi Ministry of Information, Falcon, Pompei, p. 45.
2 *Industrial Structure and Development in Saudi Arabia*, op. cit., Table 2.3.

well as a number of recently appointed technocrats.[1] And the monarch daily holds open court, access to which is available to any citizen. That the growing number of foreigners in the labour force – the projected annual rate of growth between 1975 and 1980 was 21 per cent as against one of 3.4 per cent for Saudi workers[2] – represents a threat to Islam has been reflected in a decision to stop all further immigration, to concentrate instead on the more intensive training of native workers, and increasingly to enlarge the employment opportunities for women, who are now confined to teaching, nursing and social work. Since both education and the increased employment of women are a slower process than immigration it seems likely that development as a result will be retarded. Be this as it may, the institutions of Saudi Arabia make for a measure of unity and solidarity that was absent in Iran.

Beneath the institutions, however, Saudi Arabia is undergoing the same trauma of the spirit as other Muslim countries have undergone or are undergoing. For over the centuries Islam has undergone a change. 'The Arabs . . . during the heyday of Islamic intellectual effervescence . . . acquired the Greek learning, subjected it to investigation, experimentation and expansion in such diverse fields as algebra, geometry, astronomy, navigation, chemistry and medicine and evolved the scientific principles of empiricism . . . [they] carried this new empirical scientific attitude to Spain among other places and from there this attitude burst out into Europe in the fifteenth and sixteenth centuries and the Renaissance.'[3] But at a time not exactly identified (it may have been the end of the sixteenth century) the Muslim world was cut off from the mainstream of science and technology. 'The result was that whereas Western Europe and America went forward with new advances in knowledge these communities shut themselves off and were content to dwell in a

---

1 *Saudi Arabia's Search for Security*, op. cit., p. 13
2 Ibid., p. 14.
3 *Crisis in Muslim Education*, pp. 45, 46

kind of intellectual isolation. Then when the west encroached upon them suddenly, they woke up and found themselves unable to defend either their political or their intellectual independence . . . There was a widespread suspicion that the new education which the European powers sought to bring them was calculated to destroy their cultural heritage.'[1] It might also be added that Muslims became instilled with a sense of inferiority.

The problem now is whether to remain in isolation or to endeavour to move forward. Iran has chosen the first; Saudi Arabia is still attempting the second. One cannot prophesy the outcome, for the task facing Islam is enormous – to embrace modern technology without altering its own essential nature. And the attack on the Grand Mosque at Mecca in December 1979 could be interpreted in one of two ways: as the reappearance of a Messiah, as foreseen by the Prophet, in the spirit of Christ, who received some support; or as an early sign of opposition to official Islam.

Whatever the significance of that incident, one can conclude only that Saudi Arabia barely has enough time – fifty or sixty years – to enable her to 'develop'. Her present level of oil production undertaken in deference to the needs and pressure of the outside world, may help her in her international relations, through which she aims at a settlement of the Palestinian issue; on the other hand, from the point of view of internal development she is depleting too fast and she may see the failure of her attempt to industrialise. On the face of it, a fast rate of depletion makes available increased financial resources for 'development'; but it also creates obstacles in that it aggravates the shortage of native skills, the inadequacy of the physical infrastructure, and the disparity in incomes which goes along with inflation. The choice is Saudi's own, but the industrialised countries ought not to add to her problems by seeking to denude her further of her one major resource.

---

1 Ibid., p. 52.

Saudi Arabia

Al-Farsy, Fouad, *Saudi Arabia. A case study in development*, Stacey International London, 1978.

Central Department of Statistics, Kingdom of Saudi Arabia Ministry of Finance and National Economy, *Statistical Year Book*, Fourteenth Issue 1398 AH – 1978 AD.

Dawisha, Adeed, *Saudi Arabia's Search for Security*, Adelphi Paper No. 158, International Institute for Strategic Studies, London, 1979.

Field, Michael, *A Hundred Million Dollars a Day*, Praeger, New York, 1976.

Husain, S.S., and Ashraf, S.A., *Crisis in Muslim Education*, Hodder and Stoughton, Jeddah, 1979.

Industrial Studies and Development Centre, *Industrial Structure and Development in Saudi Arabia*, Riyadh, 1977.

Middle East Economic Survey, Volume XXIII, No. 25, 7 April 1980.

Saudi Ministry of Information, *Industry*, Falcon, Pompei.

Thompson, W. C., *A Positive Attitude to New Producers*, Shell Reprographics, London, 1979.

CHAPTER 7

Kuwait: the Multi-millionaire City State

The countries adjoining Saudi Arabia on the southern shores of the Arabian (historically, and still to the Iranians, Persian) Gulf may all be described as emanations of Saudi. The indigenous Arabs are probably descendants of refugees from Arabia proper during the vicissitudes of history which that country has undergone. In spite of certain differences there are basic traits common to all the Gulf countries. The most important country in terms of oil is Kuwait.

A grey, flat desert of less than 10,000 square miles, beneath it a sea of oil, and on its surface a population of little over 1 million people, with an average income per head of $12,565 (US), as compared with $7,000 in the US and $3,500 in the UK: such is the modern city state of Kuwait. (Incidentally, it is a superb external vantage point from which to glimpse the serious tone of current Iranian television, the programmes consisting invariably of debates chaired by the ubiquitous mullah.)

The global figures are, however, deceptive. First, while the surface has been almost fully explored, the exploration has yet to probe more deeply, in the hope that reserves of gas may be found, while off-shore there has as yet been little exploration. Nor has there as yet been any secondary recovery of the known reserves beneath the solid ground. Secondly, the gross figure of population conceals the proportion of native Kuwaitis to non-Kuwaitis. 'At the end of 1978, Kuwait's population was put at just over 1 million, but the share of Kuwaiti indigenous citizens, estimated by the Planning Minister at 43 per cent, is considered on the high side.

Some figures put it as low as 33 per cent.'[1] Amongst the working force, as distinct from the population in general, the proportion of non-Kuwaitis may fairly safely be put at around 70 per cent.[2] Finally, the average figure of income per head conceals a considerable disparity between Kuwaitis and non-Kuwaitis. A survey undertaken by the Ministry of Planning on family spending in 1977 showed that 'the average middle-income Kuwaiti family spend KD (Kuwaiti Dinar) 473 a month, 28.5 per cent more than the average non-Kuwaiti family's spending of KD 368.'[3]

Kuwait entered the world of oil after the Second World War, the two main concessionaires being BP and Gulf, who formed the Kuwait Oil Company, the most important of three oil companies in the country. There is no reason to believe that Kuwait's experience with the oil companies differed from that undergone by other countries. 'The oil companies used to exploit the richest fields which in most cases required no effort in controlling reservoir pressure, i.e. no additional investments, and over-produced these fields beyond the maximum limits, leaving the smaller fields of relatively lower productivity or doing away with pay zones of relatively poor quality crudes. They also failed to apply, except on a limited scale, the secondary and tertiary recovery methods, which in turn require additional investments.'[4]

In 1968 – that is, while the companies in Kuwait were still privately owned – OPEC recommended to its member states the introduction of petroleum conservation controls. Of the seven Arab countries in OPEC only four followed this recommendation, some of the seven finding that it was already too late, the deterioration of the reservoirs having gone too far. Kuwait finally promulgated a Conservation Law in 1973. This has been described as 'the most thorough among conservation laws,'[5] even though it left several matters out of

1 Rajai M. Abu-Khadra, 'Review of the Kuwaiti Economy', *OPEC Review*, Volume III, no. 2, Summer 1979, p. 44.
2 In the 1975 census 71 per cent of total employment was non-Kuwaiti.
3 Abu-Khadra, op. cit., p. 62.
4 OAPEC, *Reservoir Engineering*, Kuwait, 1979, p. 147
5 Ibid., p. 146

account – for example, the setting of a maximum limit to production, the introduction of restrictions on gas/oil separation, conservation of flared gas, etc.

Meanwhile, Kuwait had followed other oil-producing countries in an attempt to secure a larger share of the profits. For example, there was a fifty/fifty profit sharing agreement in the 1950s, though participation in ownership did not come until twenty years later. In January 1974 the Government entered into a participation agreement by which it took ownership of 60 per cent of the Kuwait Oil Company. Then in December 1975 the Company was totally nationalised. Of the other two concessions, one stretched into Saudi Arabia and was exploited in conjunction with that country and Getty; it could not therefore be wholly nationalised. This triple agreement cannot be said to have made for good reservoir practice, and is an outstanding example of the need for unifying the control of a single field. The problem however, is small in relation to others that have been encountered.

There are at least two major factors which have impaired Kuwait's ability to produce. First, when Mossadegh nationalised the oilfields of Iran in 1951 the burden of making good the shortfall fell upon Kuwait. There ensued a rate of extraction by the companies in excess of that which would have been consistent with the maximum long-term enjoyment of the field. A field may be subject to under-production without damage to the future, but it may not be subjected to over-production without subsequent harm. Secondly, when nationalisation began to appear on the horizon, the companies, as in Venezuela, reduced their investment, though the reduction appears to have fallen more on exploration than on current extractive facilities. Table 13 shows how capital expenditure fell dramatically from 1968 to 1975, the year of nationalisation. Once nationalisation had taken place capital expenditure increased in three years from KD 1.5 million to KD 113.5 million. The lower levels of estimated expenditure in 1979 and 1980 probably reflect a need for less investment now that production targets are being reduced.

Table 13: *Capital expenditure by the Kuwait Oil Company*

	Capital expenditure in million Kuwaiti dinars
1968	16.8
1969	13.8
1970	5.8
1971	4.1
1972	1.6
1973	1.2
1974	1.1
1975	1.5
1976	59.7
1977	110.4
1978	113.5
1979(a)	46.7
1980(a)	80.1

(a) Estimated.
SOURCE: Kuwait Oil Company.

In spite of the increase in investment after nationalisation it is pertinent to ask whether Kuwait monitors the activities of her nationalised company more closely than she did those of the private companies. The answer is that she does not. In common with most other Arab countries the supervision exercised over the nationalised company is loose. Be that as it may, when nationalisation took place in December 1975, the Government decreed that production should be at the level of 2 mbd, as against a peak output of 3.3 mbd reached in 1972. Table 14 shows how production and exports have declined. Whether the figure of 2 mbd was fixed upon out of a desire to conserve or whether it was made obligatory by preceding events is not clear. The reduced level of company investment prior to nationalisation may have made some reduction in output inevitable and, in addition, production facilities may not have been properly maintained by the Companies.

The output of 2 mbd was kept up until the end of 1978. Kuwait then joined several other OPEC members in increasing output in 1979 to make good the shortfall which took place in

Iranian production after the Khomeini revolution. Towards the close of that year the Government decided to decrease the output to 1.5 mbd, a reduction which it finally made public on 1 April 1980. The reasons are fairly clear. Inflation had eroded the capital value of the revenues which had been invested abroad. Events in Iran had shown that too rapid a rate of depletion could lead to an atmosphere of speculation, to inflation, and to an enhanced disparity in incomes. And by now it was fairly clear that the price of oil was going to rise faster than other prices. These were all sound reasons for keeping more oil in the ground.

Table 14: *Kuwaiti production and exports of crude petroleum and refined products*

	Million barrels per day			
	Crude oil production	Crude oil exports	Exports of refined products	Refined exports as a percentage of total production
1960	1.7	1.5	0.1	5.5
1965	2.4	2.1	0.1	4.6
1970	3.0	2.6	0.3	14.2
1971	3.2	2.8	0.4	13.9
1972	3.3	2.9	0.2	12.2
1973	3.0	2.6	0.2	13.5
1974	2.5	2.2	0.2	14.3
1975	2.1	1.8	0.1	14.0
1976	2.1	1.8	0.3	18.7
1977	2.0	1.6	0.3	18.7
1978	2.1	1.8	0.3	14.0
1979(a)	2.5	na	na	na

na: not available.
(a) Provisional.
SOURCES: OPEC, *Annual Statistical Bulletin 1977*, pp. 23, 69, 78; United Nations, *World Energy Supplies 1950–1974*, pp. 211, 311; United Nations, *World Energy Supplies 1973–1978*, pp. 128, 161; OAPEC, *Sixth Annual Statistical Report 1977–78*, pp. 15, 95; *Petroleum Economist*, Volume XLVII, Number 3, March 1980, p. 135.

Various estimates have been made of the life of the known Kuwaiti reserves, ranging from fifty years, through eighty and one hundred, to two hundred and fifty. The very range of these estimates suggests their futility. For the life of reserves depends on many varying factors – the changing geology as extraction proceeds, the changing price, the rate of increase in the population, and the fluctuating political pressures, internal and external, which may prompt a faster or slower rate of depletion. All that can be sensibly said is that Kuwait should be able to continue to enjoy high revenues from oil for a fairly considerable time ahead.

How has the Emir, at the moment unsupported or unhindered by a Parliament, used the accruing oil revenues? He has followed the Islamic precept of caring for the needy by installing a modern welfare state on the most lavish scale. Medical care is free, education is free (except for a few private schools), air-conditioned housing is available on cheap terms (though one old gnarled inhabitant has held out and still lives in a house which is not air-conditioned). Non-Kuwaitis, however, have to pay the market price for their dwellings. One may be born in Kuwait and still be a non-Kuwaiti, even stateless, since one is not entitled to citizenship if born of non-Kuwaiti parents. Many, perhaps most, non-Kuwaitis are Palestinians, victims of a second diaspora inflicted on them by the victims of the first; and the differentiation between them and the country's citizens is potentially explosive.

To some extent the danger may be palliated by a wide-ranging system of subsidies which includes water, desalinated with the aid of associated gas; electric power, also generated with the help of associated gas; gasoline sold, as in Venezuela, well below the cost of production; and many kinds of foodstuff. True, it is an objective of the Government gradually to narrow the spread, or alternatively to reduce the level, of the subsidies. Indeed Kuwait Airways are now being charged the world price for fuel. But the reduction of subsidies is a long and laborious task. For the time being the social policies of this traditionally merchanting and trading community, dedicated to private enterprise, are the very reverse of 'self-help'. Indeed they render the Government very vulnerable to popular pressure and make all the more necessary the maintenance of a certain level of oil production. It is notable that the five-year plan 1976/77–1980/81, while it aims in broad terms at a modest rate of growth, none the less emphasises the improvement of the social services. It would be no exaggeration to say that, while the merchanting tradition of Kuwait inclines it to free enterprise, the importance of the social services impels it towards planning – that is, towards fitting the social services into a larger whole.

There has been an attempt to attain a more equal distribution, if not of income, at any rate of wealth. The Government, the owner of the land, has sold to private individuals below the hypothetical market value; and in so far as it has bought back again, it has done so at below the original selling price, thus bestowing on the original buyer a capital gain. Between 1946 and 1971 land purchase consumed as much as one quarter of total government revenues from oil. These expenditures, and those on a welfare state and subsidies, are a means by which the Government tries to distribute the benefits of oil revenues widely throughout the population.

This policy of conferring wealth on individuals should have created a class of entrepreneurs, but alas, the traditional merchant has hesitated to take to manufacturing entrepreneurship. Not that he was highly taxed – there are no taxes. He was simply averse to waiting for a long-delayed return. As a result the Government itself has entered the breach, forming with the individual a joint company. 'Altogether, there were twenty-six public and private companies in the joint sector in June 1977, with government participation in the range from 8 per cent to 80 per cent.'[1]

The philosophy underlying the joint company deserves elaboration. It may be encapsulated in two ministerial statements. The first is one made by the Minister of Finance and Oil in 1974: 'The incentive for private investment in indus-

---

1 M. W. Khouja and P. G. Sadler, *The Economy of Kuwait, Development and Role in International Finance*, Macmillan, London, 1979, p. 130.

trial and service development was still insufficient and the initial role of the joint sector system was to encourage private involvement in areas considered valuable to Kuwait's long term development but which possessed the deterrent elements of being capital intensive or having a protracted payoff.' The second statement is that made by the Prime Minister also in 1974: 'They [the government-appointed directors] should act as normal commercial directors but should direct the policy of the company in accordance with strategic goals laid down by the Government as and when found necessary.' As interpreted by Khouja and Sadler, '. . . it is recognised that the prospect of profit alone may not result in the fulfilment of some of the State's social and political aims, so that government involvement in joint enterprises will probably remain permanent in most cases and will be a further example of Kuwait's adherence to the private enterprise principle in economic activity while at the same time attempting to control the country's destiny by securing its social and economic base within the Arab world.'[1]

Government entry has over the years extended beyond manufacturing proper into, for example, hotels, banking and insurance. In addition to government participation, manufacturing enjoys some help from a unified customs duty of 4 per cent *ad valorem* on almost all imports, while some items produced locally benefit from a graduated tariff – car batteries, paints, biscuits. Other imports – asbestos, cement and welded steel pipes – are prohibited altogether. In spite of these restrictions, Kuwaiti trading policy remains broadly liberal. Yet manufacturing makes only a small contribution to the Gross Domestic Product – 3.6 per cent in 1973–74 and 5.0 per cent in 1975–76. Indeed by 1975–76 manufacturing had never exceeded 5 per cent of GDP. Table 15 shows that in the 1970s oil provided 70 per cent of GDP, with nearly all the rest of GDP coming from services of various types, covering public administration, finance and retailing.

The basic reason for the smallness of the manufacturing

1 Ibid., pp. 129–30.

sector, in spite of all the forms of help – tariffs, subsidies, even cheap loans – is the small size of the internal market. Would it be possible to enlarge the market – by, for example, collaboration with Saudi Arabia, or even by the formation of a wider combination embracing the Arab countries?

Table 15: *Sectoral contribution to Kuwaiti Gross Domestic Product*

	Percentage sectoral contribution to GDP		
	1971/72	1973/74	1975/76
Agriculture and fishing	0.3	0.2	0.3
Manufacturing	3.1	3.6	5.0
Electricity and water	3.5	2.0	2.3
Construction	3.0	1.0	0.9
Wholesale and retail trade	6.7	5.3	5.8
Transport and communications	3.0	3.1	2.6
Banking, insurance and finance	1.6	6.4	4.6
Public administration and other services	11.8	9.9	8.5
Crude oil and natural gas	67.0	68.5	70.0
Total GDP in million Kuwaiti dinars (a)	1,347	2,112	3,279

(a) Current market prices.
SOURCE: Central Statistical Office, Ministry of Planning, *Annual Statistical Abstract*, 1978, Edition XV, Table 193, p. 215.

Most businessmen and probably most members of the Government would be in favour of such an enlargement. They might even possibly prevail. Powerful arguments, however, could be mounted against them. The common external tariff of any enlarged market would need to be considerably higher than Kuwait's present moderate tariff, and Kuwait would lose what remains of her liberal trading tradition. In addition, she would be subsidising with her cheap services manufacturers needing a higher tariff than herself. In short, she could find herself, like the United

Kingdom vis-à-vis the European Community, a heavy net contributor. And like the United Kingdom she could learn that she is subject to mistrust — Saudi Arabia is horrified by the number of non-Kuwaitis.

There seems little hope of industrialisation for Kuwait, unless, after long experience with the technology of oil, now responsible for 70 per cent of the GDP, her educated people can adapt themselves to some other technology the nature of which cannot at the moment be foreseen. This will not be easy, for 'The technique(s) of production in mining and oil tend to be specific and do not lend themselves to useful application in other sectors'[1] It has been argued that from 1970 to 1975 there was some success in diversifying the economy. Many new industries were established, and since 1973 some of the older industries achieved good financial results.[2] Certainly it is the case that imports of machinery and transport equipment rose from 31 per cent of total import value in 1971 to 41 per cent in 1976 at a time when the value of total imports was itself rising rapidly. This suggests that an expanding industrial base required these imports, but to what extent that base was being established outside oil is unclear. For oil continues to stand apart from the rest of the economy, and the only industrialising route apparently open is to take oil farther downstream. Kuwait already produces urea and ammonia more cheaply than Western Europe; her exports consist of fertilisers and liquid petroleum gas as well as crude oil, and she has plans to move into aromatics and ethylene. Kuwait has perforce to move downstream, for her own internal consumption of energy is increasing at a rate of around 20 per cent a year, as compared with an annual growth rate in all Arab countries of 14.2 per cent and in developed countries of 5.6 per cent (1975–76).[3] If this rate of

---

1 Kamal S. Sayegh, *Oil and Arab Regional Development*, Praeger, New York, 1968, p. 82.
2 Yusif A. Sayigh, *The Economics of the Arab World. Development Since 1945*, Croom Helm, London, 1978, p. 105.
3 Abdulaziz Alwattari, *Downstream Developments in OAPEC countries: Refining and Petrochemicals,* OAPEC-JCCME Seminar, Tokyo, October 1979, p. 51.

growth continues, Kuwait would cease to be an exporter of crude oil around 1995.

She is likely, however, to be increasingly important as an exporter of refined products, these giving her a higher added value than exports of crude oil. While the absolute volume of crude oil exports has significantly declined, that of refined products has been broadly maintained, while the proportion of refined exports in relation to total production has steadily increased, as is shown in the final column of Table 14.

When the oil exports have gone, how does Kuwait, apart from exporting refined products, earn her living? More important still, when the oil is gone, how does she live except as a rentier state? Some provision for this eventuality was made in December, 1976, when there was created the constitutionally inviolable Fund for Future Generations, with a capital of KD (Kuwaiti dinars) 850 million ($2,900 million), to be replenished each year with 10 per cent of the State General Reserve. The fund is to be kept intact for the post-oil era. Meanwhile, the Fund apart, proceeds from the sale of crude oil and its derivatives have been invested abroad as well as at home. The income arising from them is estimated to account for 8 to 10 per cent of the Government's total budgetary revenues.[1] Table 16 shows data on Kuwait's foreign earnings and investment.

The earnings from investment income are substantial and correspond to a figure of about 10 per cent of government income, or a little more. This investment income is Kuwait's second largest earner of foreign currency after oil exports, and from 1974 to 1977 it increased by 150 per cent, while oil income increased by only 10 per cent. Government revenues have been so large in recent years that substantial sums have been allocated to reserves each year. For example, in 1976/77 41 per cent of total government expenditure was allocated to reserves and to the Kuwait Fund for Arab Economic Development – a Kuwaiti-financed institution which provides finance for projects in other Arab countries and now also other non-Arab developing countries.

1 Abu-Khadra, op. cit., p. 50.

Table 16: *International receipts and payments*

	Million Kuwaiti dinars(a)		
	1970/71	1974/75	1975/76
Exports of goods and services (f.o.b.)	645.7	2,992.5	2,650.5
Property and entrepreneurial income from the world (investment income)	102.8	232.0	330.0
Imports of goods and services (c.i.f.)	240.6	552.6	737.3
Property and entrepreneurial income to the rest of the world	212.6	512.8	106.0

(a) One Kuwaiti dinar equals approximately 3.4 US dollars.
SOURCE: Central Statistical Office, Ministry of Planning, *Annual Statistical Abstract*, 1978, Edition XV, Table 197, p. 217.

Revenues during the financial year July 1979 to June 1980 were expected to be KD 3,241 million, of which 96.3 per cent would be from oil. This level of revenue would be 41 per cent greater than the projected level for 1978/79. Yet spending was expected to rise by only 16 per cent, from KD 1,940 million to KD 2,250 million, in spite of the official rate of inflation of 22 per cent. This conservatism of Kuwaiti planners was reflected also in the development budget, which was being increased by only KD 5 million to KD 395 million. Development spending increased very rapidly in the years between 1973/74 and 1977/78, but the difficulty of spending all this money to the long-term benefit of the economy is now fully realised, and a more measured pace is being followed. However, the implication of this deceleration, particularly as revenues are rising rapidly, is either that there will be substantial surplus funds to invest abroad or that production and exports of oil will have to be reduced, or both. Surplus revenue in 1978/79 after contributions to the State General Reserve and the Reserve Fund for Future Generations was KD 600 to 700 million. In 1979/80 KD 617 million were planned for the former and 10 per cent of revenues to the latter, as usual. Thus, Kuwait would be expecting to invest abroad at least KD 1,500 million in 1979/80 (about $5,500 million). This figure is fully commensurate with foreign assets in 1979 which, according to one estimate, are put at about $22,000 million, to which might be added $4,000–$6,000 million in private hands, making a total of foreign holdings of the order of $28,000–$30,000 million.[1] Other unofficial estimates go much higher, and may indeed be accurate. It is certainly the case that these assets are increasing rapidly with the surpluses created from existing levels of revenue and expenditure.

Even the conservative estimate of around $30,000 million represents a strain on foreign stock markets. Total market capitalisation of American stock markets amounted in mid-1974 to $628,000 million, or about 56 per cent of the total market valuation of all world stock markets. Of the figure of $628,000 million, only $17,800 million or 2.8 per cent of the total market capitalisation would fall outside the restrictions imposed by the Securities Stock Exchange Act of 1934. According to this Act ownership of more than 5 per cent of the total issued shares of a public company has to be notified to the Securities and Exchange Commission; any transaction in and, *a fortiori*, the disposal of such shares is subject to restriction. Kuwait's assets thus amount to almost double the unrestricted shares in which she might choose to deal on the American stock exchanges.[2] While the restrictions of the London stock exchange are looser, there is an unwillingness by UK companies to accept Kuwaiti shareholdings. Thus it is not only the oil-producing countries which can show a lack of 'absorptive capacity' – that is, an inability to put oil monies to a tangible use; the capital markets of the world are equally limited in their capacity and willingness to absorb the financial surpluses arising from oil.

As a result, Kuwait's investments are probably only thinly

---

1 Abu-Khadra, op. cit., p. 50.
2 Hikmat Sh. Nashashibi, *Arab development . . . through cooperation and financial markets*, Alshaya, Kuwait, 1979, pp. 39, 40.

placed in portfolio stocks and probably feature more prominently in short to medium-term bonds. While the income from these investments has been maintained, their capital value has been reduced, partly because of inflation, partly because of the secular fall in the foreign exchange value of the dollar, to which the Kuwaiti dinar is tied. (Kuwait's one major portfolio investment is in Mercedes-Benz, her shareholding representing 10 to 15 per cent of the total. She has, however, no seat on the board and the investment has brought her no industrial gain.) Although Kuwait, like other Arab countries, has been frightened by the decision of the US Administration in November, 1979, to freeze Iranian assets, as creating a precedent which might in unforeseeable circumstances affect herself, she is unlikely to exchange her dollar holdings; their volume and therefore potential effect on the foreign exchange market would be too great. Faced with the inability, or unwillingness, of Western financial markets freely to absorb the surpluses arising from oil, Kuwait is seeking to turn increasingly from investing to lending. In 1978 she ranked third in the league of Eurobond lenders, as is shown in Table 17.

Kuwait's ambition would be to lend both to developed countries and to Arab countries. In the case of the latter such funds as the Arab Development Fund would finance the infrastructure, while Kuwaiti investment banks would finance what it is hoped would be viable industrial projects. In this way Kuwait would become a financial centre, the main act of investment being aided by all manner of ancillary services.

Can she realise this ambition? These are early days, but one remains a little sceptical, less perhaps of the ability to lend to developed countries than of the scope for finding suitable projects in the Arab world. If Kuwait herself has difficulty in finding such projects at home, so also must her neighbours, if each is to go, as now, its own way.

Even if Kuwait does succeed in becoming a financial centre, what would such success mean for employment? There is at present no population policy, though immigra-

Table 17: *Fixed rate Eurobonds in 1978*

	Million US dollar equivalents	Per cent
Deutsch marks	5,481	54.2
US dollars	3,093	30.6
Kuwaiti dinars	450	4.4
Dutch guilders	350	3.5
Sterling pounds	289	2.9
European units of account	177	1.7
French francs	103	1.0
Euro yen	78	0.8
Others	91	0.9

SOURCE: Kuwait International Investment Co., *Annual Report*, 1978, p. 13.

tion is not allowed in the absence of a work permit. Without a population policy, the present population of marginally over one million could well rise by the end of the century to just under three million. Could a successful financial centre provide work for such a population? It is impossible to say. But if the answer turns out to be in the negative, and if the Western world has not by then found a vent for the energies of the growing armies of educated unemployed, Kuwait will find herself face to face with a moral problem of having to cope with unemployment. One can only hope that the traditional ingenuity of the Kuwaitis will find an answer to it.

Whereas some other countries produce at or near capacity and spend the oil revenues domestically on development and current consumption, this is not possible for Kuwait. The smallness of the country places a clear limit on expenditure which can be usefully undertaken domestically. Moreover, the process of economic diversification is a long-term activity, as the Kuwaiti Minister of Oil emphasised when he said, 'there are no short cuts in economic and social development ... This process will occupy the better part of a century at the very least. For it involves not only the physical changes in the landscape, but more basically the fundamental transformation of the human being and the social fabric of the society.'¹ This recognition requires revenue over a long period, and thus favours keeping current production levels as low as possible.

In all oil-producing countries the revenues required for development and other domestic expenditures themselves require a certain level of production. However, in 1970 Kuwait was in the fortunate position of being able to provide for domestic oil consumption and finance gross fixed capital formation and development expenditures with only 58 per cent of the value of her total oil production. This figure fell to 45 per cent in 1976 and 35 per cent in 1977.² It was probably lower still in 1979. Thus, Kuwait's required production is sufficiently low that for some time ahead she should be able to pursue a firm regulation of the rate of depletion.

There may be some who will argue that Kuwait will not be able to reduce her production to 1.5 mbd in 1980 as planned, because of pressures for domestic expenditure. Certainly the present discrimination against non-Kuwaitis, added to the natural expectations of Kuwaitis so-called, is likely to compel increases in social spending if tensions within the society are to be kept at bay. The income so far generated by oil is, however, so great in relation to the size of the population that it should make possible the maintenance, and indeed the expansion of the social services, notwithstanding a declining production in oil.

Kuwait

Abu-Khadra, Rajai M., 'Review of the Kuwaiti Economy', *OPEC Review*, Volume III, No. 2, Summer 1979.

Alwattari, Abdulaziz, *Downstream developments in* OPEC *countries: Refining and Petrochemicals*, OAPEC-JCCME Seminar, Tokyo, October 1979.

Central Statistical Office, Ministry of Planning, *Annual Statistical Abstract*, 1978.

---

1 *Middle East Economic Survey*, Vol. XXII, No. 48, 17th September 1979, Supplement p. 3. A speech given by HE Shaikh Ali Khalifa al-Sabah, at the Oxford Energy Seminar, 13th September 1979.
2 *OPEC Review*, March 1979, documentation section.

Khouja, M. W. and Sadler, P. G., *The Economy of Kuwait, Development and Role in International Finance*, Macmillan, London, 1979.

Kuwait International Investment Co., *Annual Report*, 1978.

Middle East Economic Survey, Vol. XXII, No. 48, 17 September 1979.

Nashashibi, Hikmat Sh., *Arab development . . . through co-operation and financial markets*, Alshaya, Kuwait, 1979.

OAPEC, *Reservoir Engineering*, Kuwait 1979.

OPEC Review, March 1979.

Sayegh, Kamal S., *Oil and Arab Regional Development*, Praeger, New York, 1968.

Sayigh, Yusif A., *The Economics of the Arab World. Development since 1945*, Croom Helm, London, 1978.

CHAPTER 8

The Smaller States of the Arabian Gulf

The smaller States of the Arabian Gulf – the United Emirates (population of, say, one million), Bahrain (population, about 300,000), and Qatar (population, 250,000[1]) – are in many ways replicas of Kuwait. They all produce more oil than is needed for their own requirements. With one exception – the United Arab Emirates – they have all been careless of depletion. Because of the smallness of their populations they offer only a limited market for manufactured products. In each case more than half the population consists of immigrants – Arab or Asian. In Qatar and the United Arab Emirates considerably more than half the population are immigrants. All show a high rate of increase in the internal consumption of oil, particularly gasoline. They vie with one another in the attempt to become a financial centre and take the place of the defunct Beirut; and, in spite of the fragmentation, each is jealous of its sovereignty. Even the United Arab Emirates are far from being really united.

The State over which the exhaustion of oil reserves looms most closely is Bahrain, a green island while its neighbours are part of a dun desert. Table 18 gives details of Bahrain's production of and trade in crude petroleum and refined products. Production from its own Bahrain field reached a peak of 77,000 bd in 1970, and has since fallen steadily to just over 50,000 bd in 1979. There is thus an annual rate of depletion of 5 to 6 per cent. True, secondary recovery is practised. There are also experiments in tertiary recovery, and exploration for new reserves goes on. The fact remains

1 Qatar, Ministry of Information, 1980 estimate.

that the present known reserves are approaching exhaustion, and by the end of the century there may be little or no production from this field. Bahrain also has a half share with Saudi Arabia of the Abu Safa field, operated by Aramco. This management by a single operator makes for a more effective use of the field than the sharing practised between Kuwait and Saudi Arabia. Production from Abu Safa, and substantial other production from Saudi Arabia, are reflected in the rising level of crude oil imports.

Table 18: *Bahraini production and trade in crude petroleum and refined products*

	Thousand barrels per day			
	Crude oil production	Exports of crude oil	Imports of crude oil	Exports of refined products
1950	30	0	123	124
1960	45	0	161	183
1965	57	0	135	153
1970	77	0	175	206
1971	75	0	182	214
1972	70	0	166	204
1973	68	0	177	202
1974	67	0	186	207
1975	61	0	152	168
1976	58	0	161	206
1977	58	0	201	237
1978	55	0	219	238
1979*	50	0	na	na

*Provisional.
na: not available
SOURCES: OAPEC, *Annual Statistical Report*, 1976–77, pp. 15, 47; 1977–78, pp. 11, 37; UN, *World Energy Supplies, 1950–74*, pp. 210, 310; *Petroleum Economist*, January 1980, p. 6; UN, *World Energy Supplies, 1973–78*, p. 130.

All Bahrain's crude is now wholly nationalised, although until the end of 1979 the Bahrain Oil Company had 40 per cent of the equity, as against 60 per cent shared between Standard Oil of California and Texaco.

All Bahrain's crude, including the imports, is refined at the 250,000 bd refinery at Sitra, which is still owned by the US companies. This is still one of the largest refineries in the Middle East oil-producing countries. Bahrain's earnings from oil are derived entirely from the export of refined products, which have risen steadily over the years, as Table 18 shows. Despite experiments in tertiary recovery, and substantial gas reserves, the rate of depletion of Bahrain's oil means that it must consider the time when it will be wholly dependent upon oil imports, both for its domestic consumption and as a raw material for exports of refined products. The geographical proximity to and the closeness of relationship with Saudi Arabia makes the latter country the obvious source for continued and increased imports.

Qatar's endowment of oil is not much richer than that of Bahrain. Table 19 shows that a peak production of 600,000 barrels a day was reached in 1973. With the outbreak of the Egyptian-Israeli war, production was reduced by 15 per cent to marginally under 500,000 barrels a day, at which level it has continued. This volume of output cannot for long be sustained, and unless studies of new reservoirs prove fruitful, production will have to be cut, a strict policy of control over the rate of depletion having become inescapable.

The United Arab Emirates (UAE) are much better placed. They are one of the few countries with a rising ratio of reserves to production. Present production, at the rate of 1.8 mbd, represents only 60 per cent of what could be produced. It could be said, therefore, that there is a policy for the regulation of rate of extraction, and it is estimated that, at the current rate of output, the reserves at present known could last for another fifty years. Next to Saudi Arabia the UAE have the largest reserves of crude oil in the world. Table 20 gives details of production and exports. A special fund has been set up to explore further the possibilities of enhanced recovery.

Table 19: *Qatari production and exports of crude petroleum*

	Million barrels per day	
	Crude oil production	Exports of crude oil
1950	0.03	0.03
1960	0.2	0.2
1965	0.2	0.2
1970	0.4	0.3
1971	0.4	0.4
1972	0.5	0.5
1973	0.6	0.6
1974	0.5	0.5
1975	0.4	0.4
1976	0.5	0.5
1977	0.4	0.4
1978	0.5	0.5
1979	0.5	0.5

Note: There are no exports of refined products from Qatar.
SOURCES: OPEC, *Annual Statistical Bulletin*, 1977, Table 19, p. 26; Ministry of Finance and Petroleum, Qatar, *Oil Industry in Qatar 1976*, Table IV, p. 44; Qatar Petroleum Producing Authority, *Qatar Oil Offshore*, 1980.

There is an important difference in institutional arrangements between Bahrain and Qatar, on the one hand, and the UAE, on the other. In Bahrain and Qatar oil has been fully nationalised, and foreign expertise is tapped through the placing with international companies of contracts for technical assistance. In Qatar the Government entered into equity participation with the companies in 1973, when it took over 25 per cent of Qatar Petroleum Company (a subsidiary of British Petroleum) and Shell Qatar. The Qatar General Petroleum Corporation was established in 1974 to take over the Government's interest in the oil fields, and then in 1976 and 1977 QGPC gained 100 per cent control of Qatar's oil.

In the UAE, by contrast, there is only one nationalised company – in Dubai. Abu Dhabi, the desert sand whistling through its brand new streets, owns the largest reservoirs in the Emirates, but has not yet fully nationalised. Abu Dhabi joined Saudi Arabia, Kuwait, and Qatar in signing a participation agreement in December 1972, which from the beginning of 1973 gave these countries a 25 per cent equity stake in the oil companies operating in their territories. The Abu Dhabi National Oil Company (ADNOC), which had been formed in November 1971, took responsibility for Abu Dhabi's equity stake. Then in September 1974 Abu Dhabi followed Kuwait and secured a 60 per cent equity participation, effective from January 1 1974. However, Abu Dhabi has not followed other oil-producing countries in securing 100 per cent ownership. The multi-national oil companies have been able to retain a 40 per cent equity holding, although Abu Dhabi intends eventually to secure 100 per cent nationalisation. In the other oil-producing Emirates, Dubai apart, the State has a 60 per cent holding, while the companies own 40 per cent. This latter arrangement, by which the foreign oil companies retain an equity holding, and which is now unique, can give rise to strain. The international company is interested in a reasonably early payback; the State, in the shape of the Emirate, can wait for a longer time. An example of the possible tension between the two points of view can be seen in the case of a company called DMA (Abu Dhabi Marine Areas Limited). The State wished to explore an area known as the Upper Zakum; the main participating company – BP – was reluctant; its participation was thereupon transferred to a Japanese company; in the event substantial reserves of oil were found.

There is a further problem which arises no matter whether the international company enjoys a participation in the equity or operates under a contract for technical assistance – namely, whether it is ready to transmit expertise to the indigenous people, and whether, if it is, there are enough native engineers to absorb it. It has been seen that the international companies, when they had concessions, were neglectful of the development of the country in which they were working. Are they ready now, from a different standpoint, to transmit

Table 20: *United Arab Emirates production and exports of crude petroleum*

	Million barrels per day	
	Crude oil production	Exports of crude oil
1962	0.01	0.01
1965	0.3	0.3
1970	0.8	0.8
1971	1.1	1.1
1972	1.2	1.2
1973	1.5	1.5
1974	1.7	1.7
1975	1.7	1.7
1976	1.9	1.9
1977	2.0	2.0
1978	1.8	1.8
1979★	1.8	na

★Provisional.
na: not available.
Note: There are no exports of refined products from the UAE.
SOURCES: OPEC, *Annual Statistical Bulletin*, 1977, Table 21, p. 28; UAE, *Oil Statistical Review*, 1979, pp. 32, 61; *Petroleum Economist*, Volume XLVII, Number 3, March 1980, p. 135.

knowledge? It has been said that 'there is little native technological participation in the planning and execution of projects.'[1] Unless there is a willingness to transfer technological and other technical aid, the arrangement by which an international company has a share of the equity, and possibly, that by which it operates under a service contract, cannot last. Indeed, although the transfer of ownership to the oil-producing country by that very act raises substantially the income which it receives from oil, it is by no means necessarily the case that the oil company fails to earn a significant profit from the transaction. The terms of the service contract for Qatar, for example, enable the companies

---

1 A. B. Zahlan, *Transfer of Technology and Change in the Arab World*, Oxford, Pergamon Press, 1979, p. 17.

to receive back their costs in full, plus a fee of about 15 per cent a barrel.[1] This is still a highly satisfactory arrangement for the companies.

The fact that the different Gulf States are endowed with oil to different degrees also presents a problem of policy. A leaf might be taken out of the book of the International Energy Agency (the consumers' OPEC), one of the rules of which prescribes that a country with plenty of oil should help a neighbour suffering from a shortage. The corollary on the side of the producer countries would require that a State with considerable reserves still intact should help an adjoining State the reserves of which are approaching exhaustion. No such rule, however, exists. That is a measure of the looseness of OPEC as an organisation. OAPEC would undoubtedly bless such an arrangement. The reaction of the individual sheikhdoms, however, has not in the past been favourable, though it was seen above how Bahrain relies on substantial crude oil imports from Saudi Arabia, which will become of increasing importance as its own reserves dwindle.

More important than oil is gas, and in each of the three States under consideration the gas reserves exceed the oil reserves. In the United Arab Emirates it is estimated that the gas reserves can last one hundred years beyond the exhaustion of the oil reserves. And in Bahrain the drilling of certain gas reserves is being deferred. The significance of gas is that, as a form of energy, it is the key to any kind of industrialisation. Even though the gas may now be flared and therefore wasted, it has been argued that its use in the Gulf for industrial purposes 'cannot be as profitable to host governments as the production and export of crude oil itself.'[2] The argument is based on the fact that the gas is expensive both to collect and to distribute, whether by pipeline or by ship. Be this as it may, the countries of the Gulf wish to industrialise, for two good reasons. First, without industrialisation they see their one capital asset, oil, being drained from them;

---

1 *Middle East Economic Development,* Special Report, November 1979, 'Qatar', p. 13.
2 Thomas R. Stauffer, Energy – *Intensive Industrialisation in the Arabian/Persian Gulf,* Tehran, 1975, p. 45.

hence the expression '*naft* and *shaft*'. Secondly, industrialisation is a prerequisite for the defence of the insecure frontiers of the Arab homeland. It is customary to claim that the frontiers of Israel are insecure; given, however, the technological superiority of Israel, coupled with the ill-defined, perhaps undefined, nature of Israeli ambitions, from the Arab point of view it is the Arab frontiers which are insecure.

The use of gas for industry is also, it is argued, indirectly a means of conserving oil in the ground. For whatever be the product into the making of which the gas enters, if the gas finds an export market there is rendered unnecessary the use of any other form of energy, including, for example, fuel oil. The world market for energy is one, and gas for certain uses is an alternative to oil. The increased use of gas can thus contribute to the regulated depletion of oil.

For what forms of industry, however, is the gas best used? It has been estimated that the most rewarding uses of gas in the Gulf are for the refining of oil, for the energy costs of refining are high; for gas liquefaction; for the manufacture of urea for use as a fertiliser; and for the production of aluminium. 'Gulf-based plants in these industries could potentially have an important competitive edge . . .'[1] As for steel, this has been described as 'the least lucrative of the industrialisation opportunities.'[2] On the other hand, it could be argued that steel is one of the most suitable industries for an industrialising country, since it requires other things to be made for it, and other industries use it; in other words, it can create industries both in front of it and behind it.

How does this ideal set of industries compare with what has been done? Bahrain has a plant for the liquefaction of associated gas, a petro-chemical plant, and another is projected in conjunction with Saudi Arabia. It also has an aluminium plant and a ship repairing yard. The United Arab Emirates have a petro-chemical plant, an aluminium plant, and a dry dock at Dubai, and a number of establishments serving construction in the internal market. Finally, Qatar, as

1 Thomas R. Stauffer, op. cit., p. 45.
2 Thomas R. Stauffer, op. cit., p. 30.

well as constructing a petro-chemical plant, has a steel plant, opened in April 1978 and protected by a tariff of 20 per cent. The plant is a subject of controversy, it being contended that the gas is most profitably used for export in liquid form. However, steel does take the country away from oil. All the projects listed above have been in existence for too short a time to allow of a fair judgement on any of them. It can, however, be said that there is no contradiction between the manufacture of petro-chemicals and the export of crude oil since the proportion of crude oil needed for the manufacture of petro-chemicals is small. Further, insofar as the petro-chemical produced is urea as a fertiliser, there should be an export market both in non-oil producing countries in the Arab world, and in Africa. It has been calculated that 'only 4.0 kg of fertiliser per person are used in Arab agriculture, compared to 73.5 kg, 48.8 kg, and 83.6 kg per person in the United States, Europe and Australia respectively.'[1]

Bahrain is the least rich of the Gulf oil states, and has to face sooner than any other the prospect of using its oil revenues to secure an income to replace oil when it is exhausted. Yet despite never having had a level of oil production to compare with other producers, Bahrain now has one of the most diversified and sophisticated economies of the region, and probably has one of the better chances of succeeding without oil. This may be partly because its production has a longer history than in most other countries, going back to the 1930s. In addition, however, oil has never dominated the economy to the same extent as in Kuwait, Saudi Arabia, or the Emirates. Although at its peak it was 60 per cent of the GNP, the proportion has now fallen to 40 per cent, and seems certain to fall further, compared with over 70 per cent in Kuwait. Consequently, Bahrain has had to ensure that its economy does not rely on oil alone.

Bahrain's diversification has depended on four main industries. First, the liquefaction of gas; secondly, offshore banking (she has some 120 offshore banks); thirdly, the ship repair

1 Adnan Mustafa, *The Status of Arab Nuclear Potential*, given at the First Arab Energy Conference, 4–8 March 1979, p. 20.

yard; and fourthly, the aluminium smelter. As mentioned above, all the Gulf states have been trying to take Beirut's place as a financial centre. Bahrain has had more success than most countries, her great advantage being the excellence of her communications. However, she is now facing greater competition, and it is by no means certain that in this respect she has a future, especially as other countries have more revenue of their own to place in the banking system.

The Arab Shipbuilding and Repair Yard Company (ASRY) was established in Bahrain with OAPEC funds. It is jointly owned by several Arab countries, with Saudi Arabia, Kuwait, United Arab Emirates, Qatar and Bahrain having the larger shares. ASRY is now making profits, but its future is still in doubt because of the lack of cooperation among Gulf states. In spite (or perhaps because) of OAPEC's backing for ASRY, Dubai went ahead and built its own dry dock complex, which now has three docks to Bahrain's one. In the early part of 1980 none of the three docks had any work. This is but one example of the difficulty of economic integration even between economies uneconomic in themselves. Indeed, the Qatar National Navigation and Transport Company has also established a floating dock at Umm Said, which is likely to aggravate the excess capacity. ASRY has the further advantage for Bahrain of training a labour force, and helping to stimulate local industries.

Aluminium Bahrain is also now achieving some success and making money, though the raw materials are imported. Its production in 1980 was around 80,000 tons a year and the estimated projected capacity is 165,000 tons a year. It is also the intention to move downwards to rolled products. However, it will also face competition from Dubai Aluminium, which has recently been completed, and both may be rendered unprofitable. It might have been far better if some of Saudi Arabia's associated gas could have been used for a joint Saudi-Bahrain aluminium plant, rather than Bahrain's unassociated gas which could be left in the ground to an even greater extent than now.

However, it would not be accurate to say that there is no

co-operation. The supply of oil from Saudi Arabia to Bahrain has already been mentioned. Further, a joint petro-chemical plant is envisaged and is due to produce in 1983–84. In addition, about 10 per cent of Bahraini Government revenues in 1979 were derived from grants from Saudi Arabia. More importantly, however, Saudi Arabia is financing the $1,000 million causeway which will link it to Bahrain, and a decision was aimed for 1980. Bahrain is hoping that the completion of the long-awaited causeway will give a boost to its economy, which is quite possible. Finally, in spite of the presence of immigrants Bahrain has no social problem; indeed it could be described as a middle-class community.

Unlike Bahrain, both Qatar and even more so the UAE have substantial surpluses of current oil revenues over current expenditure. The UAE, in particular, have so much oil that they can afford calmly to contemplate their industrial future, and the scholarly Minister for Oil of Abu Dhabi, Dr Mana Saeed Al-Otaibi, while still busy, can retire of an evening to his estate and indulge his taste for falconry. The geological surveys of the UAE so far show that construction materials, such as gypsum, are available in some quantity, though whether they could ever form the basis for an export industry is doubtful. Industrial development would require a further examination both of local resources and of regional needs, and the fact that the examination would require time would not matter. Meanwhile, present ambitions appear to be confined to becoming, like Kuwait, a rentier state.

The projects mentioned as being undertaken by the smaller Gulf states, have more properly to be viewed against a wider background – namely, 'the inappropriate technological environment'[1] of the Arab world. In the case of petro-chemical plants it is estimated that they 'cost 50 per cent to 100 per cent more than if executed in Europe or the US. The elevated prices are due to the need to import labour, supplies and equipment and the generally weak technological base of these countries.'[2] As for steel, 'the pillar of industrialisation', the

---

1 A. B. Zahlan, op. cit., p. 17.
2 A. B. Zahlan, op. cit., p. 15.

'manpower training requirement is even more staggering than the investment: an average of 10 per cent of the cost of the project (as compared with 2 per cent in industrial countries) must be devoted to this purpose.'[1] Further, 'a study of steel development in many countries would indicate that a gestation period of about ten years is usually required for absorbing and developing the required technological know-how and skills, as well as for the development of other inter-related sectors.'[2] As for petro-chemicals, 'despite impressive looking plans . . . the Middle East will not in this category be participating on a significant scale.'[3] In sum, 'the problem facing the Arab world remains one of how much technical education and R and D is necessary in order to support policies and projects rather than one of whether to have such activities at all.'[4]

These views suggest that it will be a long time before exports alternative to crude oil can be developed. Theoretically, it may be true that the use of gas for industrial development may be consistent with the conservation of more oil in the ground. In practice, however, the long time required for industrial development means that meanwhile the oil is being drained away. Nor is the development to be accelerated by the formation of a Gulf 'common market'. There appears, in spite of the model of the Andean Pact, no disposition to form one, though the Gulf as a whole would seem a natural single entity. Even so, such a market would still remain small and in any case the requirement is that the oil-producing countries should find export markets outside themselves rather than amongst themselves.

Perhaps more important than the formation of a common market is cooperation in joint projects on the Andean model. There are now four such projects in being; all have been sponsored or instigated by OAPEC, but the participants are nation-states. The four are:

1 Omar Grine, *Transfer of Technology in the Arab Steel Industry*, in A. B. Zahlan, op. cit., p. 451.
2 Omar Grine, op. cit., p. 452.
3 A. B. Zahlan, op. cit., p. 15.
4 A. B. Zahlan, op. cit., p. 18.

1. Arab Maritime Petroleum Company (AMPTC), which has been joined by all but one of the ten members of OAPEC.
2. Arab Shipbuilding and Repair Yard (ASRY), which has been referred to earlier and which three members of OAPEC have refrained from joining.
3. Arab Petroleum Investments Corporation (APICORP), which has been joined by all members of OAPEC.
4. Arab Petroleum Services Company (APSC), which has also been joined by all members of OAPEC.

It is a pity that only one of these projects lies outside oil – namely, ASRY. The terms of reference are, however, wide and go into the field of engineering – to construct docks, centres for manufacturing sheets and pipes, workshops for machines, welding, electricity and propellors, training centres, warehouses, bases for building platforms, etc. There are also further joint projects in prospect, in particular an engineering centre in Baghad and a training centre.

These joint companies of Arab States are 'gaining a position comparable to that of international holding companies or multi-national corporations which, for long periods, were able to extend and thereby exert great control over petroleum activities in the Arab world.'[1] They are, therefore, a new development of the organisational form of the corporation and as such are unique. Nor need the shareholding be confined to governments; private investors from Arab States may participate, provided that governments retain a majority holding. It is desirable that more of these companies should be established and that their purview should extend beyond petroleum, though such an extension may require an amendment of the terms of reference of OAPEC. It is clear that OAPEC has been the engine; that being so, its powers should probably be enhanced. The formation of companies encompassing more than one Arab state requires a core of experts, both if the activity is confined to petroleum, and *a fortiori* if it extends beyond it. Such experts

1 *Petroleum and Arab Economic Development,* OAPEC, Kuwait, 1978, p. 159.

at the moment are few, and their paucity is an obstacle to the receipt of technology. 'The number of practising Arab reservoir engineers with a minimum experience of five to ten years is small. In Iraq, for instance, they are not more than eight, in Egypt they are fewer than ten (including university professors), in Kuwait they are about ten, in Libya they are fewer than five.'[1] The solution of this problem requires initially a central training, research and information collecting centre. It is difficult to see how this centre can be other than under OAPEC; it is interesting that ASRY has attached to it a training centre. When the initial centre has acquired strength, ancillary centres could be established in the member countries.

The lesson for Arab countries is that they should combine both in the formulation of joint projects – including those away from oil – and in the necessary training for successful execution of the projects. Otherwise, the Arab homeland will remain a collection of so many sheikhdoms, each incapable of industrial development on its own, each seeing its oil being sucked away, and joint effective defence against the imperialist Israel receding into the never-never. Industrial development is all the more imperative in that in the case of Bahrain and Qatar, the oil is near exhaustion; the UAE alone have the reserves and, therefore, the time to reflect.

The Smaller States of the Arabian Gulf

Al-Otaiba, Mana Saeed, *Petroleum and the Economy of the United Arab Emirates*, Croom Helm, London, 1977.

Mallakh, Ragaei El, *Qatar: Development of an Oil Economy*, Croom Helm, London, 1979.

Middle East Economic Development, Special Report, 'Qatar', November 1979.

Mustafa, Adnan, *The Status of Arab Nuclear Potential*, First Arab Energy Conference, 4–8 March 1979.

OAPEC, *Petroleum and Arab Economic Development*, Kuwait, 1978.

Qatar Petroleum Producing Authority, *Qatar Oil Offshore*, 1980.

Stauffer, Thomas R., *Energy – Intensive Industrialisation in the*

---

[1] *Reservoir Engineering*, Kuwait, 1979, p. 18.

Arabian/Persian Gulf, Conference on the Persian Gulf and the Indian Ocean in International Politics, Tehran, 25–27 March 1975.

United Arab Emirates, *Oil Statistical Review*, 1979.

Zahlan, A. B., *Transfer of Technology and Change in the Arab World*, Oxford, Pergamon Press, 1979.

CHAPTER 9

Iraq and Algeria: the Revolutionary Regimes

Islam stands neither for capitalism nor for Communism. 'In a capitalist economy attention is focussed on the material and technical problem of maximum production, exploitative capitalism and imperialism, and is strongly biased against a social and moral approach. Similarly, the Marxist approach replaces the market by the State as the mechanism for choice and makes man a passive agent not to be trusted with freedom. Neither of these, therefore, is Islamic in nature.'[1]

The Gulf States so far dealt with – Saudi Arabia, Kuwait, Bahrain, Qatar and the United Arab Emirates – may all be described as 'private enterprise' countries. In other words, the ruling regimes would prefer that economic activity be privately undertaken. This, however, is more easily said than contrived. The traditional merchanting mentality in the Gulf States looks for a quick financial return; manufacturing, by contrast, requires that the return on an investment be awaited for a longer spell. In practice, therefore, the activity directly undertaken by the State tends to be much greater than that which one would normally associate with a 'private enterprise' economy. Generally speaking, small-scale industry is left to the private sector, while the important industries are run by the State.

As a result, while there is a theoretical distinction between different Islamic states in their political ideology, the practical difference is not all that great. The greatest difference probably concerns the degree of economic inequality, this being less in the two nominally Socialist countries – Iraq and Algeria – than elsewhere.

1 S. S. Husain and S. A. Ashraf, *Crisis in Muslim Education*, pp. 81, 82.

IRAQ

Iraq, ruled by the Ba'th Socialist Party, is avowedly a Socialist, but not (though helped and influenced by the Soviet Union) a Communist country. Private activity is confined to the consumer goods industries, and a private company is limited in the amount of capital which it may use – 70,000 Iraqi dinars. The oil industry was nationalised in 1972. There are, however, no service contracts with the international companies, the Government company itself both producing and distributing; and the number of foreign experts employed is few.

The first oil concession was granted in 1925 to the Anglo-French company, Turkish Petroleum Co, renamed in 1929, Iraq Petroleum Co [IPC]. The Kirkuk field was discovered in 1927 and production began in 1928. It is now clear that in the following years oil discoveries in the Middle East were so prolific that the oil companies often found it desirable to hold back development to prevent a fall in price. 'The IPC quickly acquired a reputation for dragging its feet in the development of its oil concession.'[1] For example, it delayed the pipeline through Syria and delayed development of the Qatar concession. This policy annoyed the Iraqi Government, for, after negotiation to receive royalty payments and quantities of free oil, the receipts were limited by action outside Iraqi control. Although the oil-producing countries now accuse the companies of thereby exploiting them, in retrospect this policy was a form of depletion control, for it helped in the longer term to conserve oil and keep prices and thus revenues higher than they would otherwise have been.

Throughout the 1930s and 1940s the Iraqi Government expressed its dissatisfaction with the royalty system and with the level of revenue it was receiving. The 50-50 profit-sharing agreement had begun in Venezuela in 1948 and had spread to Saudi Arabia in 1950. In 1952 Iraq reached agreement with IPC on a similar arrangement. However, as IPC was

---

1 Edith and E. F. Penrose, *Iraq: International Relations and National Development*, Ernest Benn, London, 1978, p. 73.

a subsidiary especially set up by the majors it earned no profits. Consequently, the system of 'posted prices' was evolved so that profits could be attributed to the production of Iraqi crude oil and thus taxed. The 'posted prices' and, therefore, the Government revenues based on them were, however, effectively under company control.

Following the revolution of July 1958, in which the monarchy was overthrown, the new Government took a more determined stance in securing control over Iraqi oil. The fate of Mossadegh induced some caution, although in 1952 and 1953 Iraqi production had been substantially increased to take advantage of reduced Iranian production, to the benefit both of the companies and of the government. However, in December 1961 the dramatic Law 80 was passed, which expropriated 99.5 per cent of IPC's original claim without compensation. Normally when a concession is granted the treaty contains a clause stipulating that if part of the conceded area continues unexploited, the State may regain possession of the unexploited part. The concession treaties signed by Iraq contained, however, no such clause. Yet it was found that the concessionaires were exploiting little more than one per cent of the ceded territories, which covered almost the entire country. The State thereupon laid its hand on the unexploited areas.

Iraq was to suffer for Law 80 over the next ten years as the oil companies took their by now familiar action of favouring other countries in their investment and production policies. Between 1962 and 1970 the companies increased production in Iraq at an annual rate of 4.7 per cent, as compared with 13 per cent elsewhere in the Middle East and 11.1 per cent in Iraq in the period before Law 80, from 1952 to 1961.[1] Iraq was thereby deprived of additional revenue, for in a period when the posted price was not rising the only means of increasing revenue was to increase production. Similarly, capital investment by IPC fell dramatically after 1961. Its investment in Iraq was 22.8 million Iraqi dinars in 1960, 22.5 million in 1961,

1 Iraq National Oil Company, *The Nationalisation of Iraq Petroleum Company's Operations in Iraq. The Facts and the Causes,* 1973, pp. 8, 9.

Oil: The Missed Opportunity

but a mere 4.7 million in 1962, and from 1963 to 1969 it averaged only 0.7 million a year. The extent of this reduction was stunning.[1]

Nationalisation, announced on June 1, 1972, should have come therefore as no surprise. Yet the concessionaire companies resisted it. By March 1, 1973, 'Iraq had already marketed 48 million tons of oil out of a maximum pumping capacity of 58 million tons, whereas before nationalisation the companies had laid down a production schedule of a maximum of 31 million tons on the pretext that there were difficulties in marketing and other technical problems. The difference between 31 and 48 million tons is very revealing and indicates whether there were or were not any serious difficulties.'[2] Faced with these facts the companies, nine months after the Act of Nationalisation, 'bowed down and accepted the form of agreement.'[3]

In nationalising the oil industry the Iraqi Government, unlike other governments in the Middle East, seems to have had more in mind than just the capture for the State of oil revenues. It appears to have seen it as a step towards Pan-Arabism and the effective defence of the Arab homeland. The land of Babylon is not without ambitions. 'We don't look on this piece of land, here in Iraq, as the ultimate limit of our struggle. It is part of a larger area and broader aims, the area of the Arab homeland and the aims of the Arab struggle.'[4] 'Whoever enjoys effective influence in the Middle East will be able to influence Europe and Japan . . . when the sources of oil become independent, Japan will be able to deal with the Arabs directly . . .'[5]

Be this as it may, since nationalisation output has been more than doubled, reaching 3.4 mbd in 1979, as against 1.5 mbd in 1972, as can be seen from Table 21; while reserves have been doubled – they probably amount to around 35

---

1 Ibid, p. 11.
2 Saddam Hussein, *Current Events in Iraq*, Longman, London, 1977, pp. 59-60.
3 Ibid., p. 59.
4 Saddam Hussein, *Social and Foreign Affairs in Iraq*, Croom Helm, London, 1979, p. 69.
5 Ibid., p. 99.

billion barrels. Iraq has been very keen to expand oil production. From 1975 to 1978 expansion was hampered by the slow growth in world demand for oil, but Iraq quickly took advantage of the fall in Iranian production in 1979. Iraq has, as a result, become second only to Saudi Arabia in production. Even so, the ratio of reserves to output is declining, and exports at the end of the century are unlikely to exceed 4 mbd.

Table 21: *Iraqi production and exports of crude petroleum and refined products*

	Million barrels per day	
	Crude oil production	Exports of crude oil and refined products(a)
1950	0.1	0.1
1955	0.7	0.6
1960	1.0	0.9
1965	1.3	1.2
1970	1.5	1.5
1971	1.7	1.6
1972	1.5	1.4
1973	2.0	1.9
1974	2.0	1.9
1975	2.3	2.1
1976	2.4	2.2
1977	2.5	2.3
1978	2.6	2.4
1979	3.4	na

na: not available.
(a) Iraq's exports of refined products are negligible.
SOURCES: OPEC, *Annual Statistical Bulletin*, 1977, pp. 22, 81; OAPEC, *Sixth Annual Statistical Report 1977–1978*, p. 15; United Nations, *World Energy Supplies 1950–74*, p. 210; *Petroleum Economist*, XLVII, Number 3, March 1980, p. 135.

The output of 3.4 mbd in 1979 fell somewhat short of capacity, which is currently estimated to be 4 mbd. This shortfall appears to be the result not so much of a deliberate policy of regulating the rate of depletion as of the integration of oil within the total planning system. Most of the oil-producing countries have followed the example of the oil companies in regarding oil as something apart. Iraq, however, has made oil part of the whole planning system with the result that if, for example, emphasis is laid in the plan on nuclear power, so much more oil is conserved in the ground. The act of conservation is, however, indirect rather than direct.

The plan consists of three parts: a twenty-year qualitative or descriptive plan; a five-year quantitative plan; and an annual plan consisting of a list of projects. The aim of the plan is not to maximise oil revenues but to maximise the Gross Domestic Product. In constant 1975 prices, this rose from 5,135 million Iraqi dinars in 1977 to 5,763 million Iraqi dinars in 1978, an increase of 12.2 per cent.[1]

It is not easy to obtain a picture of the industrial progress of Iraq, largely because the gross figures given above are not broken down in recent official statistical publications. Revolutionary regimes are, after all, given to secrecy, and perhaps the factories installed with Russian aid were not working all that well. Moreover, the available statistics are often not comparable, and vary considerably between sources. Nevertheless, it is clear that manufacturing provides only a small fraction of GDP, with great reliance being placed on oil, as would be expected, and also a large service sector. This is borne out by official data on employment in 1977. Table 22 shows that, of the total economically active population, 45 per cent were employed in services. Moreover, of the total service employment of 1,414,161 nearly one million were involved in community, social and personal services. This distribution of the working population may reflect an emphasis on the social services.

Although manufacturing undoubtedly remains a small part of GDP, it has not been neglected. From a value of 100 in 1962, an index of industrial production rose to 156 in 1970

1 Republic of Iraq, Central Statistical Organisation, *Annual Abstract of Statistics* 1978, p. 134.

and 375 in 1977.[1] Most manufacturing output, nevertheless, is related either to oil or to construction, with very little engineering capacity. For example, the Basrah refinery is being expanded, and as well as the chemical fertiliser plant at Khor al-Zubeir, a petro-chemical plant is being built at Shuaiba. Like other states in the Middle East, Iraq is embarking on an iron and steel complex, but it will face the same problems of whether this will be profitable and can find a market. However, it is perhaps not surprising that Iraq's major effort is concentrated on maintaining and expanding its oil production capacity.

Table 22: *Distribution in Iraq of economically active population by field of economic activity*

	1977
Agriculture, hunting, forestry, fishing	943,890
Mining, including oil	36,835
Manufacturing	284,395
Building and construction	321,696
Finance, insurance, real estate and business services	31,089
Electricity, water and gas	23,190
Wholesale and retail trade, restaurants and hotels	224,104
Transport, storage and communications	177,799
Community, social and personal services	957,979
Unknown	58,237
Unemployed	74,725
Total economically active population	3,133,939

SOURCE: Republic of Iraq, Central Statistical Organisation, *Annual Abstract of Statistics 1978*, pp. 38, 39.

Because agriculture is such a major employer it remains a crucial part of the economy, as the palms along the Tigris testify. But as in many other developing countries, particularly those with substantial sources of income from sectors

---

[1] Economist Intelligence Unit, *Quarterly Economic Review of Iraq*, Annual Supplement 1979, p. 6.

other than agriculture such as oil, agriculture suffers from the migration of workers from the countryside to the towns. The Iraqi Government has tried to introduce socialist policies, which in agriculture has meant State farms, collectives and centralised co-operatives. However, agriculture in Iraq would appear weak both in these reformed areas and in conventional areas where there is little use of crop rotation, irrigation or adequate marketing.[1] This weakness is reflected not only in the large fluctuations in agricultural output from year to year, but also in the fact that agricultural production was 17 per cent lower in the period 1973 to 1978 than in the preceding period 1967 to 1972.[2] Moreover, net imports of foodstuffs continue at a high level, representing over 10 per cent of total imports.

For the rest, one has to rely on other than statistical evidence. In an interview given in January 1977 Saddam Hussein, a President of resolution and great perceptiveness of mind, said: 'We have not given the social services first priority until the beginning of this year. During the past three years we gave priority to investments in industrial and agricultural development.'[3] The emphasis on investment may have meant austerity in consumption; there are plenty of fish in the Tigris, but there was never any at the Hotel Baghdad.

It is known that among the industrial developments are those for petro-chemicals, nuclear power and steel. And the investment figures for agriculture should probably be doubled, since it is there that the private sector is strongest. As for the social services expenditure, between 1969 and 1976 it grew 10 times – from 11.8 million Iraqi dinars in 1976.

Given the absence of more detailed and reliably consistent figures, and the unreliability of impressions based on necessarily narrow visits, one is left with the image of a country

---

1 Yusif A. Sayigh, *The Economics of the Arab World. Development since 1945*, Croom Helm, London, 1978, p. 35.
2 Annual Abstract of Statistics 1978, op. cit., p. 54.
3 Saddam Hussein, *Social and Foreign Affairs in Iraq*, Croom Helm, London, 1979, p. 2.

which, in spite of a turbulent past, has a determined leadership which aims at the headship of the Arab world. In that aim it can count on the solidarity of a country in which income and wealth are fairly evenly spread. Nonetheless, it suffers from the same defect as the rest of the Arab world – the lack of an initial industrial base and of trained leaders of the second rank. In spite of its industrial ambitions it is not a country which is likely to be profligate with its oil reserves, even though it joined in with Saudi Arabia, Kuwait and others in making good the shortfall in oil supplies which resulted from the Iranian revolution in 1978. 'Our oil supplies may be summed up in the slogan that one of the last two barrels produced in the world must come from Iraq. What I mean is that we should not use up our oil wealth too early, but also that the very last barrel should not be an Iraqi barrel, because the world, by then, will have abandoned this source of energy for a new one.'[1]

In its efforts to deplete its oil resources prudently, Iraq, like other oil-producing states, has been greatly helped by the rise in oil prices in the past decade. For example, in 1970 Iraq would have required 122 per cent of its then production to secure enough revenue to finance development expenditure, gross fixed capital formation, and cover domestic oil consumption. By 1977 this figure had fallen to 91 per cent, and was doubtless lower still in 1980.[2] Consequently, Iraq now has greater freedom of action to pursue the regulation of depletion, although this has not yet emerged as firm government policy.

Given the absence of detailed figures, it is difficult to form a judgement about the 'development' potential of Iraq. In spite of the attempt to complement oil with nuclear power, one is left with the view that oil, as a springboard for industrialisation, is not enough. If this proves to be the case, there will need to be a more determined control of the rate of depletion of oil, this being the only, if not the main, source of revenue as far ahead as the outsider can now foresee.

1 Saddam Hussein, *Current Events in Iraq,* Longman, London, 1977, p. 58.
2 *OPEC Review,* March 1979, documentation section.

ALGERIA

The commercial exploitation of oil in Algeria, as distinct from its discovery, coincided with the beginnings of the Algerian war for independence from France – 1955–56. Long after the war for independence had been won, the war over oil continued, France resorting to ruse after ruse to retain control over the oil. The net result was diminished investment, an undue exploitation of the wells, and a legacy to an independent Algeria more exiguous than it need have been.

The French Government, having during the Second World War lacked oil, established on its morrow, à la mode française, a Bureau de Recherches Pétrolières, with the double object of either searching directly for oil within France and the territories regarded as belonging to her, or indirectly through chosen companies. The main area of search was the Sahara, a search which was astonishingly successful.

While the war for Algerian liberation was still being waged, negotiations were entered into over the future of the Sahara. The first proposal made by the French Government was that Algeria should obtain independence but that the Sahara should remain French. This proposal was rejected, on the ground that the Sahara might constitute a base for a new colonial empire. The second proposal was that the Sahara should be regarded as an 'internal sea', serving, through its wealth in oil, all adjacent States. This proposal too was refused. Finally, it was proposed that the Sahara should be subject to a Franco-Algerian condominium. This in turn received the answer 'No'. It was estimated by President Boumédienne, the first President of an independent Algeria, that these efforts by France to retain the Sahara in one form or another prolonged the war of independence by at least two years.[1]

The Algerian war of independence was finally concluded by the Evian treaties of 1962. But as far as oil was concerned the war went on. There were at least three aspects of the treaties touching on oil which were to the advantage of

1 Rabah Mahiout, *Le Pétrole Algérien,* E. and A. P., Algiers, 1974, p. 117.

France, not of Algeria. First, there was established a 50-50 Franco-Algerian organisation to exploit the subterranean Sahara. The composition of this organisation implied that any initiative by Algeria could be blocked by France. Secondly, French companies were given priority over the companies of any other country. Thirdly, the profits on which the French companies were taxed in Algeria were the result of prices declared by them and need not, therefore, be prices actually charged; the taxable profits could, therefore, be understated. These one-sided arrangements could not objectively be expected to last, and the French companies showed that they knew it. They exploited to the full existing reservoirs and undertook little further exploration.[1] 'They limited themselves to the maintenance of existing facilities and, as far as exploration was concerned, to the minimum of their contractual agreements . . . This reduction in investment quickly brought in its train a fall in production.'[2]

Faced with this situation the Algerian Government reacted with surprising caution. In 1963 it constituted its own organisation SONATRACH (Société Nationale de Transports et de Commercialisation de Hydrocarbures) for the task of transporting and selling oil, clearly a limited activity. It was not until three years later, 1966, that the functions of SONATRACH were enlarged to encompass exploration and production. In the meantime, the French companies continued to enjoy a privileged position, while undertaking to help with the industrialisation of Algeria by distilling oil and gas into fractions of higher added value. In fact, little appears to have been done, the clause relating to industrialisation remaining a 'dead letter'.[3] Small wonder that in the 1970s and early 1980s Algeria became one of the most aggressive of OPEC members.

Throughout the 1960s, the Algerian Government took a number of steps to gain greater control over oil production, including the nationalisation of non-French companies. By

1 Ibid. pp. 122, 123.
2 *Dix Ans d'Efforts*, Ministére pour la Planification, Algiers, 1975, p. 156.
3 Raba Mahiout, op. cit., p. 132.

the end of 1969, Government control over production was only 25 per cent, but Algeria needed more revenues for the ambitious 1970–73 Plan. The war between France and Algeria over oil finally reached its climax in 1971, when the Algerian Government took over 51 per cent of the shares of the French companies operating in Algeria, expropriating the remaining 49 per cent, and nationalising all gas deposits as well as all land pipelines conducting either oil or gas. It was added that France would be supplied with Algerian oil at the world price. The general French reaction was the withdrawal of all engineers and technicians from the oil wells. SONATRACH, which seems to have been prepared for this eventuality, then stepped in, providing, insofar as it could, the requisite personnel. One particular French company, ELF-ERAP, went further and stopped all payment of taxes to the Algerian Government.

Matters could clearly not remain in this impasse. Later in 1971 an agreement was reached according to which the Algerian Government paid compensation for the expropriated 49 per cent of the shares, which were vested, not in the original parent company, the Compagnie Française des Pétroles (CFP), but in a subsidiary company registered under Algerian law, Total-Algérie, which received its appropriate share of the oil produced in the relevant wells. Specific amounts by way of investment were required of Total-Algérie, which in return received certain benefits if the amounts specified were exceeded – an arrangement not dissimilar from that between the Saudi Arabian Government and Aramco. Roughly similar agreements were reached with the other main French company, ELF-ERAP. Algeria had taken steps to gain control over her oil resources 'which have proved more daring, quicker and more successful than any other oil country in the Arab world.'[1]

In spite of the agreements, relations between Algeria and France remained strained, complicated by the presence in France of Algerians considered by Algeria to be ill-treated,

---

1 Yusif A. Sayigh, *The Economics of the Arab World*, Croom Helm, 1978, p. 551

Iraq and Algeria: the Revolutionary Regimes

and the continued presence in Algeria of some French. Algiers, rising steeply above the sea, still wears the French appearance which it had during the Second World War. Indeed there is a deliberate attempt to Arabise the country. The war over oil cannot be said to have been definitively ended until 1974, when ELF-ERAP agreed to extend its cooperation with SONATRACH into new fields and CFP agreed to join with SONATRACH in off-shore exploration.

In sum, the war over oil, having lasted over ten years after the war of national independence, left Algeria, through SONATRACH, with control over both the oil and the gas reserves. The estimates of reserves vary, with OPEC giving a figure for crude oil reserves in 1979 of 6.3 billion barrels, compared with other estimates of 9.0 billion barrels. An official Algerian figure is not made public, but an estimate of between 8.0 and 10.0 billion barrels would seem reasonable. Gas reserves are substantially larger than oil reserves, standing perhaps at two and a half to three times the oil reserves in barrels of oil equivalent. One reputable source puts proven, probable, and possible gas reserves at 24 billion barrels of oil equivalent.[1]

The relative sizes of these reserves means that Algeria is relying heavily on gas for her future export earnings and as the source of finance for industrialisation and development. Table 23 shows that oil production has been proceeding at a rate of just over one million barrels per day. The lifetime of the oil reserves can thus be estimated at around 20 years – a short time. Moreover, water and gas injection now has to be used to maintain pressure and production. 1979 oil production was below the level of 1978, and plans to reduce production in 1980 by between 5 and 10 per cent have been made public. By the late 1980s production would at most be 1.0 mbd, and could be as low as 0.8 mbd.

By contrast, commercial production of natural gas has risen rapidly from its beginning in 1961 at 231 million cubic

---

1 *OPEC Oil Report*, Second Edition, 1979, Petroleum Economist, p. 179. The Economist Intelligence Unit suggests a figure nearly twice as large as being more reasonable. In the view of the author this would be too high an estimate.

metres to 14.1 billion in 1978 and 17 billion in 1979. In the past fifteen years Algeria has taken a leading role in developing large-scale LNG export projects, and she is now much involved in establishing natural gas exports. Following the failure by the United States' Government to approve gas export deals, Algeria is concentrating her gas exports on Western Europe. She expects, probably correctly, that Western Europe will require much increased natural gas imports during the 1980s, and thus she is relying on two natural gas pipelines: one to Italy via Tunisia, and one directly to Spain. SONATRACH is trying to secure an increase in the price of gas, for this is more important for Algeria's future than increases in the price of oil. Algeria's future gas production will thus depend in part on its price, but the 'decision is not simply a question of price; it is, as with other producers, a question of our access to long-term capital, to markets for the output from our new factories, of significant progress in achieving a measure of transfer of technology, and of a sense that our partners in Europe (or the US) are sympathetic to our long-term goals.'[1]

This greater awareness of the benefits from exploiting oil and gas more slowly is reflected in the decision of the Algerian Central Committee in December 1979 to slow down the industrialisation policy, and abandon the crash programme to exploit the country's oil and gas reserves over the next fifteen years. This, however, has not always been the approach to depletion and industrialisation. In the light of the reduction in oil production during the independence war, one would have expected a more careful depletion policy. The effect of the reduction in output appears to have been the reverse – namely, to cause the Algerian authorities to exploit the reserves to the full with the purpose of accelerating industrialisation. Industrially, the French regime had left Algeria poor, with just a few, mainly construction, industries, catering for the home market – admittedly, larger than elsewhere, with a population of around 20 million people.

1 Nordine Ait-Laoussine, 'Developments in the Natural Gas Industry of Algeria', *OPEC Review*, Vol. III, No. 2, Summer 1979, p.9.

Table 23: *Algerian production and exports of crude petroleum and refined products*

	Million barrels per day	
	Crude oil production	Exports of crude oil (a)
1960	0.2	0.2
1965	0.6	0.5
1970	1.0	0.9
1971	0.8	0.7
1972	1.1	1.0
1973	1.1	1.0
1974	1.0	0.9
1975	1.0	0.9
1976	1.1	1.0
1977	1.2	1.1
1978	1.2	1.1
1979	1.1	na

na: not available.
(a) Includes exports of refined products, which are less than 100,000 bd.
SOURCES: OPEC, *Annual Statistical Bulletin 1977*, pp. 14, 82; *Petroleum Economist*, Volume XLVII, Number 3, March 1980, p. 135; OAPEC, *Sixth Annual Statistical Report 1977–78*, pp. 11, 15, 46; UN, *World Energy Supplies 1950–1974*, pp. 194, 281.

Export industry was there none, with the exception of agriculture.

Before independence, agriculture was the largest sector of GDP, along with services, each contributing about 30 per cent. Food and wine were also important exports. However, while agriculture is still easily the biggest employer, and supports about half of the population, Table 24 shows that in 1977 it represented only 7.1 per cent of GDP. Moreover, agriculture no longer provides exports but, in fact, substantial quantities of food requirements have to be met by imports. Despite attempts to favour agriculture, Algeria still has to import a fifth of her cereal consumption, one-third of milk and almost all butter and sugar. This decline in agriculture is reflected in the fact that from 1965 to

1975 oil exports rose from 50 per cent of export earnings to 95 per cent.

Table 24 also shows that manufacturing industry remains relatively small, only 10.3 per cent of GDP. Before independence, manufacturing consisted largely of food processing, textiles, cigarettes and clothing. It is true that since then there has been diversification into iron and steel, petro-chemical production and fertilisers, and many thousands of jobs have been created. Nevertheless, oil and services continue to dominate the economy, and there is no easy means of altering this pattern. Since 1965, industry in Algeria has become increasingly centralised, as the State has nationalised foreign-owned companies. Indeed, economic growth is being sustained through the rapidly increasing public sector investment. Gross fixed capital formation, as a share of GNP, rose from 43.1 per cent in 1974 to 48.1 per cent in 1977, and 54.9 per cent in 1978.[1]

Table 24: *Structure of Gross Domestic Product in Algeria in 1977*

	1977	
	Billion dinars	Per cent of total
Agriculture, forestry and fishing	5.7	7.1
Oil and gas	22.3	27.7
Mining and power (other than oil)	1.3	1.6
Building and public works	10.8	13.4
Industry	8.3	10.3
Transport, trade and services	27.2	33.7
Import duties and taxes	5.0	6.2
GDP at market prices	80.6	100.0

SOURCE: Economist Intelligence Unit, *Quarterly Economic Review of Algeria,* Annual Supplement 1979, p. 7.

Algerian industrial output, placed roughly in order of importance, would appear to be as follows: the export of liquefied gas; the distillation of oil and gas down-

---

1 *Arab Economist,* March 1979, Volume XI, Number 114, p. 29.

stream into, for example, plastics and, possibly, synthetic rubber – Algeria appears to have moved farther downstream than other oil-producing countries; and the beginning of industries, such as steel, produced by traditional methods, to cater for the internal market, as well as other industries, producing mainly for the domestic consumer, which are unlikely to have any export potential. There still, however, appears to be no real engineering industry – the car industry, for example, being an assembling operation for imported components. The large industries belong to the State; the smaller industries are left in private, and more frequently than not, in foreign hands. The nationalised industries are an interesting experiment in workers' participation, the workers being represented on the boards of directors and meeting in assemblies to define policy and the apportionment of expenditure over the coming year.[1]

How, in the light of this industrial profile, does the industrial future of Algeria look? If, in combination with the French companies, she can develop further her distillation of hydrocarbons, her industrial exports could grow, indeed, to the point of enabling her to do to gas what she has so far confined to oil – keep some locked in the ground. This, however, is merely to raise a larger question – namely, whether an economy based almost entirely on petro-chemicals is sufficient unto itself, or whether more is required to render the economy really industrialised? This is a large question, affecting all the purely oil-producing countries, and the answer for the time being is best left in abeyance.

Be it noted meanwhile that this chapter has dealt with two Socialist countries – Iraq and Algeria. It might have been thought that Socialist countries would be more prudent in their use of their oil and gas reserves than countries more oriented towards 'private enterprise'. The descriptions of the two countries given in this chapter show that that is not necessarily the case. Common to both countries is a degree of secretiveness; this common feature apart, Iraq has shown

---

1 Mostefa Boutefnouchet, *Le Socialisme dans l'Entreprise,* E. and A. P., Algiers, 1978, pp. 43-45.

itself to be a more provident user than Algeria and this difference is to be explained not by any difference in regime, but in the history which preceded 'Socialisation'.

Iraq and Algeria

Ait-Laoussine, Nordin, 'Developments in the Natural Gas Industry of Algeria', OPEC Review, Volume III, No. 2, Summer 1979.

Arab Economist, March 1979, Volume XI, Number 114.

Central Statistical Organisation, Republic of Iraq, *Annual Abstract of Statistics*, 1978.

Economist Intelligence Unit, *Quarterly Economic Review of Algeria*, Annual Supplement 1979.

Economist Intelligence Unit, *Quarterly Economic Review of Iraq*, Annual Supplement 1979.

Hussein, Saddam, *Current Events in Iraq*, Longman, London, 1977.

Hussein, Saddam, *Social and Foreign Affairs in Iraq*, Croom Helm, London, 1979.

Iraq National Oil Company, *The Nationalisation of Iraq Petroleum Company's Operations in Iraq. The Facts and the Causes*, 1973.

Mahiout, Rabah, *Le Petrole Algérien*, E and A. P., Algiers, 1974.

OPEC Review, March 1979.

Penrose, Edith and E. F.,*Iraq: International Relations and National Development*, Ernest Benn, London, 1978.

Dix Ans d'Effort, Ministère pour la Planification, Algiers, 1975.

Mostefa Boutefnouchet, *Le Socialisme dans l'Entreprise,* E. and A. P., Algiers, 1979, pp. 43-45.

CHAPTER 10

The United Kingdom: the Odd Man Out

There are basically two questions to be examined in considering the problem of North Sea oil. Did the procedures adopted by the United Kingdom enable it to secure as much as possible of the economic rent? Does the policy of the United Kingdom enable the Government to exercise an effective control over the rate of depletion? The answer to the first question is certainly in the negative; the answer to the second is probably so.

The procedures adopted to secure the economic rent have varied with time, ending in a system of taxation which must be one of the clumsiest and most complex in the world.

The sub-soil resources in the United Kingdom belong to the State, and there are two possible procedures for allocating their exploitation to private companies. First, the licences may be sold by auction to the highest bidder, the bidding taking the form either of a cash payment or of royalty payments offered to the Government (either way it is still possible for the Government later to impose further conditions on the licensees). The second method of allocating licences is according to the discretion of the Government, with the Government again laying down certain conditions.

There is considerable controversy as to which method is the better. The argument of traditional economists is that auctioning the licences secures the economically most advantageous result. It is contended that, through an auction of the licences, the most efficient company will be chosen; that company will presumably have the lowest costs and thus will be able to bid most, while also being able to extract the oil most efficiently. Further, it is argued that the auction of licences enables the owner of the oil, in this case the Government, to capture the economic rent. The economic rent on an oil field was defined in Chapter One as the difference between the market value of that oil and the costs of production including an acceptable level of profit. This level of profit should be high enough to compensate for the risk entailed in the exploration of oil, but not more than this. Insofar as it is above this level, there is a 'rent' which the State is entitled to recoup, generally by means of taxation. The difficulty is that neither the company nor the Government can determine in advance what the appropriate level of either rent or taxation should be.[1]

A discretionary system, by contrast, allocates licences according to administrative or political criteria, normally for a nominal rent. Consequently, the Government must later, once oil has been found, raise taxes to enable it to capture the economic rent.

The distinction between the auctioning method of allocating licences and the discretionary method is in fact unreal. Throughout history, the initial payment, even in a bid, has been shown to be insufficient, and the State has had no other option but progressively to raise its taxes. For example, the most profitable field found in the North Sea has been the Forties Field. The immediate acquisition by the Government of the economic rent would have entailed an initial bid of several thousand million pounds sterling – something scarcely conceivable.

The British Government, in its first and subsequent allocations of licences, has adopted in the main the discretionary method. It sought to secure two aims: first, to favour British companies like BP; secondly, to accelerate exploitation. In the first aim it failed, some 60 per cent of the concessions in the North Sea having now gone to American companies. In the second aim – quick exploitation – it was successful. 'Once the oil was discovered, both the speed at which the major finds

---

[1] For a discussion of this theory and the practice of licence allocation in the United Kingdom and Norway see K. W. Dam, *Oil Resources: Who Gets What How?*, University of Chicago Press, London, 1976.

were recorded and the oil companies went ahead, was quite extraordinary by comparison with anywhere else in the world.'[1]

Table 25 shows that, with production beginning in 1975, self-sufficiency was likely to have been reached in 1980, and, on the assumption that oil consumption continued to be between 1.8 and 2.2 mbd between 1980 and 2000, could be maintained until the end of the century.[2] The level of production through the 1980s is uncertain, for it depends not only on new discoveries and developments, but also on new techniques which may increase the rate of recovery from fields already producing. The 1979 Department of Energy estimates for reserves on the United Kingdom Continental Shelf vary between a conservative 17 billion barrels and a speculative 32 billion barrels.[3] Both these estimates depend on new discoveries and development, which may be deterred by higher taxes. On the other hand, they may be accelerated by the prospect of higher prices. The lower estimate reflects the deterrent view taken of high taxes; the higher, by contrast, tends to reflect the view taken on the basis of the expected higher prices. Even the lower figure for reserves would permit production of over 2 mbd between now and 2000, while the maximum figure would make possible a higher rate of production. From the point of view, then, of securing rapid production the licencing policy followed would appear to have been successful. The same cannot be said of the attempt to secure for the State the maximum revenue.

The first licencing round resulted in the award of fifty-three licences involving fifty-one companies, and covering 348 blocks of an approximate area of 32,000 square miles. The second round in 1965 and the third round in 1969 differed from the first round only in being smaller and, significantly, in favouring applicants willing to accept par-

---

[1] Adrian Hamilton, *North Sea Impact. Off-shore Oil and the British Economy*, International Institute for Economic Research, London 1978, p. 11.
[2] HMSO, *Energy Policy. A Consultative Document*, February 1978, p. 100.
[3] HMSO, *Development of the oil and gas resources of the United Kingdom 1979*, p. 4.

Table 25: *Oil production from the United Kingdom Continental Shelf 1975–1983*

	Million barrels per day								
	1975	1976	1977	1978	1979	1980	1981	1982	1983
Actual production	0.0	0.2	0.7	1.1	1.6(a)				
Forecast production						1.4–1.6	1.7–2.1	1.9–2.3	2.3–2.8	2.3–2.8

(a) 1979 average January to October.
SOURCES: HMSO, *Development of the oil and gas resources of the United Kingdom, 1979*. pp. 3, 38; *Petroleum Economist,* Vol. XLVI, No. 1, January 1980, p. 39.

ticipation in control, though not necessarily in the equity, by a State company. A niche, therefore, had been opened for a possible State company.

This niche was closed again when the Conservative Government came to power in 1970 and embarked on a larger fourth licencing round in 1971/72. In particular, no State company could play a part where it had no participation in the equity. This round also saw the UK's only experiment in the auctioning of licences. Of the 436 licences on offer, fifteen were put up for tender, with discouraging results.

From the second round in 1965, the number of exploration and appraisal wells drilled fell from an average of forty-nine a year from 1967 to 1969, to twenty-four in 1970 and twenty-eight in 1971. This fall could conceivably have been due to the provision relating to State participation. But it could also have been due to pessimism over the prospects for significant oil discoveries. This interpretation seems to have been borne out by the big increase in drilling activity from 1972 onwards after the discovery of many important fields (Auk, Brent, Argyll, Beryl, Piper, Dunlin, Thistle and Heather were all discovered between 1971 and 1973).

The Labour Government of 1974 renewed State participation. In the fifth licencing round of 1976/77 and the sixth round of 1978/79, all licencees were required to accept at least 51 per cent equity participation by the British National Oil

Corporation (BNOC). BNOC's share in North Sea oil probably amounts to 20–25 per cent. Provision was also made for BNOC to play a part in the management of consortia which had previously obtained licences and in which BNOC had no equity.

The pattern was thus developing on the same lines as in the Middle East, the Government trying to gain both an increased share of the economic return from North Sea oil, and greater control over oil policy. The evolution of OPEC had shown that at a stage in the development of an oil-producing country the oil companies would be more in need of supplies of crude oil than the country would be of the oil companies' expertise. Consequently, the Government would be able to insist on stricter terms for participation and on higher taxation. 'The oil companies would be locked into their enormously expensive North Sea wells and pipelines, and they would not find it easy to replace the crude oil elsewhere if they decided to write off those investments in the face of governmental insistence on greater participation or a greater governmental take.'[1]

In 1979 there was a further change of government followed by a further change of policy. It was announced that in a seventh round of licencing there would be no privileges for BNOC, no automatic equity in new licences and no control where there was no equity. What are likely to be the consequences of this policy?

Even before the policy of 1980, it is not clear that British Governments were successful in securing a fair share of the economic rent as defined in this book. The 1973 report of the Public Accounts Committee of the House of Commons was very critical of Government efforts to tax North Sea oil.[2] It pointed out that corporation tax was totally inadequate for this purpose, as companies were able to set off against North Sea profits not only capital expenditure in the North Sea, but also tax losses accumulated by companies within their world-wide activities. The UK subsidiaries of the multi-national

---

1 Dam, op. cit., p. 19
2 House of Commons, *Report of the Committee of Public Accounts 1972/73*.

companies, when buying oil from their parent companies, were charged the posted price applicable to other subsidiaries of the same companies, say, in the Middle East. This price was higher than the market price, and thus the profits and tax liability in the UK were kept to a minimum. Between 1965 and 1973 the oil majors' corporation tax liability in the UK was £500,000. It is hard to believe that the profits made did not warrant a higher tax payment. Consequently, the 1974 White Paper made it clear that taxes on North Sea oil would be raised, loopholes to avoid payment of corporation tax closed and State participation introduced.[1]

The State was faced with a number of options to capture the economic rent, once it had decided on a discretionary system to allocate licences. The Government could impose an 'excess profits tax' on North Sea activities, an output tax such as royalties based on the value of sales or the volume of production, or gain direct access to the economic rent through nationalisation, or through participation of less than 100 per cent. In the event, the Government decided on a complex system which included all three.

The tax system was based on royalties, corporation tax, and a new petroleum revenue tax (PRT), and a 'ring fence' around North Sea activities. First a royalty of 12½ per cent is charged on gross revenue. Second, a petroleum revenue tax is charged on assessable profits on each oilfield (more specifically, on the assessable profits of a participator in an oilfield) after allowing as expenses operating costs, royalties and 175 per cent of capital expenditure (reduced to 135 per cent in 1979). Third, corporation tax is imposable on North Sea activities, after the deductions named plus royalties and PRT. Finally, losses incurred outside the North Sea may not be offset against tax liability on North Sea activities. The system appeared to close the loopholes identified by the Public Accounts Committee. But did it secure for the Government the economic rent?

Table 26 (ii) shows UK Treasury data on actual taxes received from 1975 to 1979, with an estimate for 1979/1980.

1 HMSO, *UK Offshore Oil and Gas Policy*, July 1974, Cmnd. 5696.

The United Kingdom: the Odd Man Out

It can be seen that even in the financial year 1979/1980 the tax receipts were not very high. However, from 1980 the revenues are expected to build up rapidly to levels which would compare with the receipts from VAT or income tax.

Table 26: *United Kingdom Government revenues from North Sea oil and gas*

(i) NIER estimates: current prices, £ million

	1975	1976	1977	1978	1979	1980	1981	1982	1983	1985
Assumptions: Price of North Sea oil $/barrel	11.9	13.0	14.3	13.9	20.3	27.9	32.0	37.9	43.3	56.8
Exchange rate $/£	2.22	1.81	1.75	1.92	2.13	2.17	2.17	2.17	2.17	2.17
Oil and gas taxes(a)		96	175	435	1,246	2,645	3,973	5,945	10,327	12,923 19,799
British Gas Corporation 'excess' return (b)	–	–	674	596	322	1,258	1,022	1,169	1,144	1,722

(a) Accruals corresponding to each period's production; payments are made later.
(b) Return after allowing for costs of conversion to natural gas and 10 per cent return on capital.
SOURCE: *National Institute Economic Review*, No. 90, November 1979, p. 58, Table 11.

(ii) UK Treasury data: current prices, £ million

	Total 1970/71 to 1975/76	1976/77	1977/78	1978/79	Total to date	Estimate 1979/80
Royalties	65	71	228	281	652	520
PRT	–	–	–	183	449	730
Corporation tax (a)	23	10	10	50	93	140
Total	88	81	238	521	1,194	1,390

(a) The corporation tax shown is the estimated proportion attributed to North Sea oil and gas.
SOURCE: *Economic Progress Report*, No. 112, August 1979, Table 2, p.3.

The reason for this is that, while profits on some fields have already been important, tax liability other than royalties has been avoided because of generous allowances, mainly for capital expenditure. As these allowances are exhausted government receipts will rise quickly.

The Government is reluctant to make projections of tax receipts from North Sea oil in the mid-1980s. The level of these revenues depends crucially on the price of oil, which is internationally determined. When the price of oil rises it is apparent that the gross revenues from oil sales will rise, and so will the value of the Government's receipts from these sales. But the revenues will also be affected by the level of the exchange rate, since the price of oil is denominated in dollars. For example, if the price were $30 per barrel and the exchange rate $3.00 to the pound sterling, the sterling value of the receipts from the barrel would be 30/3, i.e. £10. Should the pound depreciate to an exchange value of $2.00, the sterling value of the receipts would have increased to 30/2, i.e. £15. It follows that, when the value of the sterling in terms of the dollar increases, as it did in 1979–80, the sterling value of revenues from the North Sea are adversely affected and government receipts are lower than they otherwise would have been. The high value maintained for sterling has thus had a detrimental effect on Exchequer receipts.

The Treasury claimed at one stage that the Government take from the various taxes would be over 70 per cent of net revenues. The figure of 70 per cent has, however, been challenged. 'It is customary for new oil countries to start the ball rolling by offering a generous tax regime, and then tighten it up as prospects improve as a result of discoveries and developments. The UK is no exception to this rule'[1] In the period up to 1985 the Government's share of net revenue (i.e. after capital and operating costs) is estimated by the National Institute of Economic and Social Research (NIESR) to be between 55 and 65 per cent, depending on the value of sterling and the price of oil.[2] A not dissimilar view has placed the government share of net revenue on selected fields at between 62 and 67 per cent.[3]

The suggestion that the oil companies have done very well

---

1 Christopher Johnson, *The Improvement of the North Sea Tax System*, Institute of Fiscal Studies Conference, June 5, 1979, p. 7. See also Christopher Johnson, *North Sea Energy Wealth 1965-85*, Vols. I and II, Financial Times, London, 1978.
2 S. A. B. Page, 'The value and distribution of the benefits of North Sea oil and gas, 1970-1985', *National Institute Economic Review*, No. 82, November 1977, p. 49.
3 Colin Robinson and Jon Morgan, *North Sea Oil in the Future. Economic Analysis and Government Policy*, Macmillan, 1978, Table 5.1, p. 96.

out of North Sea oil is supported by many writers. 'The companies press their luck all the time, and so far they can hardly believe what they are getting away with in the North Sea.'[1] The companies have managed to persuade the Government that they require a rate of return far greater than that achieved in other parts of the economy. 'It is clear that North Sea income from 1976 to the present day, including both company and government income, does not yet cover capital expenditure from 1973 to date; even by 1985, the North Sea company sector may be showing only a very small cumulative surplus. But the normal industrial and commercial company sector, in the UK and other industrial countries in most years, runs a financial deficit as it borrows to finance expansion, so the North Sea companies will be doing relatively well to get into surplus.'[2]

There appears, therefore, to be a fair unanimity of view that the Government was not securing from the North Sea the full economic rent. The Government accepts that this indeed has been the case, as its subsequent action reveals. Initially the petroleum revenue tax was a flat rate of 45 per cent. In 1979 the Government, in an attempt to raise more revenue from the North Sea, raised the petroleum revenue tax to 60 per cent and reduced the capital allowances which could be set against profits from 175 per cent to 135 per cent of total capital expenditure. The effect of both measures – the increase in the petroleum tax and the reduction in the capital allowances – was to raise the government take. In the Budget of March 1980 petroleum revenue tax was further raised to 70 per cent, and from 1981 the oil companies were to be obliged to make an advance payment on their PRT liabilities. The crucial question is whether this increase in taxation has led the companies to deplete their fields more quickly than they otherwise would have done and whether it has acted as a disincentive to further exploration and development.

No firm answer can be given about existing fields, but since prices have been rising at a faster rate than taxes it

1 Guy Arnold, *Britain's Oil,* Hamish Hamilton, London 1978, p. 92.
2 Johnson, op. cit., p. 2a.

may be assumed that the rate of depletion has not been accelerated. But what about the fields to come, particularly the marginal fields still awaiting exploration? A tax system with high marginal rates might discourage the development of such fields, particularly if it were thought that prices were not going to rise sufficiently.

As most if not all major fields have probably already been discovered, these new fields will be relatively small, with recoverable reserves of 150 to 300 million barrels, or even less, compared with 1,000 million barrels for Ninian, and over 1,500 million barrels for Forties. Some of these smaller fields may be liable to tax rates of over 90 per cent. Would such a tax rate deter their exploration by private companies? The companies are bound to argue that it would. The truth, however, is that nobody, not the Government nor the companies, knows in advance what would and would not deter. Only events can tell. A better judgment could, however, be formed with the help of BNOC.

BNOC was established under the Petroleum and Submarine Pipe-lines Act 1975, and began operating in January 1976. It is a totally state-owned corporation, and, according to the statute, is to be involved in the exploration, development, production, transport and refining of petroleum in the United Kingdom. It is permitted to undertake such activities abroad only with the consent of the Secretary of State for Energy. If private oil companies are genuinely discouraged by higher tax rates BNOC could undertake an exploration programme. Once finds had been made, it would be possible for BNOC to invite private company participation. Private companies are anxious to secure access to as much crude oil as possible. Once they saw that BNOC was making finds they would be likely to start bidding for licences once again. The result of this policy would be both to capture the economic rent from the more profitable fields and to secure exploration and development of the smaller, possibly marginal, fields, which are essential to maintain UK self-sufficiency into the 1980s.

It has been seen that in other countries, e.g. Canada, the

The United Kingdom: the Odd Man Out

Government has established a State company. The purpose has been three-fold: first, to enlarge the State's share of the proceeds; secondly, to gain greater physical control over the oil; thirdly, and more importantly, to enable the State to obtain greater knowledge. In Iraq, for instance, it is contended that information about the country's oil reserves is now much greater than was ever gleaned from the private companies under a participation regime. Similarly, in Canada one of the primary reasons for creating a state-owned company was a desire to obtain full knowledge of possible reserves in the frozen North and a fear that private companies would hold back from adequate exploration.

Against this history the United Kingdom is unique, having begun to travel along the same route as the United Arab Emirates and Canada in establishing a State company, and then having gone into retreat by curtailing the Corporation's activities and powers. No reason for the retreat has been given and it can only be ascribed to a belief, founded or unfounded, that a private company can operate as effectively in the public interest as in the private interest.

The intention of the Labour Government of 1974 to 1979 was that in licences issued after 1974 BNOC (alone or together with the British Gas Corporation) would have a 51 per cent equity share, with the private company as a co-licencee facing all the exploration costs, but with BNOC paying its full share of future costs once a find had been made. Norway, Iran and other OPEC members have used the same concept, and in the case of Norway, not even a State share of up to 85 per cent (held by Statoil) has deterred private companies from continuing to bid for licences.

The struggle by BNOC to secure 'participation' in licences issued under the first four rounds met determined and widespread opposition from the companies. By the end of 1975 only four small companies had accepted the 'participation' offer, because they needed Government help to raise money. The companies manifested this resistance in the same way as they had practised with such success in OPEC countries. Exploration and appraisal drilling was much reduced in

1976 compared with 1975–76 wells drilled compared with 116 in 1975. The Government was not in a strong position because it relied on the private companies to develop North Sea oil. Had BNOC been established some years earlier, then the Government's position would have been much stronger. As a result, by early 1977 all that BNOC had achieved from the few agreements reached with the companies was a right to purchase at market prices 51 per cent of their output. The Government failed to secure any equity participation by BNOC in the first four licence rounds.

Nevertheless, through the purchasing agreement BNOC became a major trader in oil, and as a result of the acquisition of a share of the equity in later consortia, it will soon be an important exploration and operating company in its own right. By the mid-1980s it is likely that BNOC will be among the top four producers of North Sea crude. Even without any equity ownership BNOC was able to sit on operating committees which manage fields, pipelines and terminals. This enabled it to offer informed advice to the Government. However, since the advent of the Conservative Government in 1979, BNOC has relinquished this right, except of course where it has an equity interest. Further, in the seventh round of licencing in 1980, BNOC will not have a 51 per cent participation in all licences, but will merely be allowed to bid like any company. It is clear that the enlargement of BNOC as an operating company is to be discouraged. Thus the Government is relinquishing a share of the ownership and thus a share of the economic rent; where BNOC has had no ownership it will also have relinquished access to the knowledge which BNOC provided by virtue of its seat on an operating committee. And the ability of BNOC to pioneer the exploration of marginal fields will have been weakened.

All BNOC profits, and all royalties, rents and licence fees paid by any company are deposited in the National Oil Account held by the Bank of England. BNOC has access to this Account for investment and development expenditure only so long as it can persuade the Secretary of State that the funds are needed. The Conservative Government which came into office in 1979 is proposing to remove BNOC's privileged access to the National Oil Account, although this will presumably also involve releasing BNOC from its obligation to pay its profits into the Account at a time when it could become a net contributor. What effect the change will have on BNOC's role of initiating is unclear, and will depend on the Minister, but the removal of a privileged access to the National Oil Account suggests that the enlargement of the role of BNOC as an operator will not be encouraged.

Can the State, without the help of BNOC, control the rate of depletion? The State certainly has powers. Under the Petroleum and Submarine Pipe-line Act of 1975, it has powers to delay developments and new investment approvals, order a reduction in output, and limit gas flaring. The limits to these powers were laid out in a statement to Parliament by the Secretary of State for Energy in December 1974.[1] There were to be no delays in the development of finds made before the end of 1975. On these finds there would be no cutbacks in production until 1982 or four years after the start of production, whichever was later. There would be no cutbacks on then existing licences until 150 per cent of the capital investment had been recovered, and cutbacks in general would not exceed 20 per cent. These are substantial powers and should in theory enable the Government to make the reductions in output of between 10 and 20 per cent which may be necessary to extend self-sufficiency through the 1980s. The question is whether the Government could impose the powers which it enjoys in theory without a state-owned company.

A policy for controlling the rate of depletion entails the reduction of production below what it would otherwise have been. This reduction can take place at one or more of three stages: first, through delayed licensing; secondly, through delayed development of fields. Each of these methods has its

1 HMSO, *Development of the Oil and Gas Resources of the United Kingdom, 1979*, Appendix 15.

drawbacks. A delay in the issue of licences and thus in exploration restricts the knowledge of reserves, knowledge which is essential for a policy to control the rate of depletion. The delayed development of fields already in production would cause a sharp drop in their output in the late 1980s. The only other option would appear to be to impose a reduction in output on fields which will be in production from 1981 onwards.

Could the Government, without detailed knowledge of the individual field, curtail output? Not without a close monitoring of activities in each field. It was seen in the case of Saudi Arabia that both Petromin and the Department of Energy closely monitored the activities of Aramco. Together they had the expertise to regulate the rate of depletion. Without BNOC the monitoring has to be done by the Department of Energy alone. Has it the expertise? It is the universal experience of the oil companies that, when they first negotiated their contracts with the UK Department of Energy, they found the Department grossly ignorant. Since those early days the Department has been advised by the BNOC as an operating company. If the advice of the BNOC is withdrawn, would the Department now have mounted sufficient expertise to enable it, in curtailing output, to have regard to the differing long-term characteristics of different fields? The answer is uncertain, though history would incline it to the negative. On the face of it, the presence of a state-owned company is essential to the effective regulation of the rate of depletion, more particularly since, as was shown in the case of the United Arab Emirates, a private company is tempted to deplete quickly. It is possible that the UK Atomic Energy Authority, which acts as an adviser to the Government on general energy questions as well as on nuclear power, has developed techniques which give it the knowledge of the oilfields needed to control the rate of production. In that case the Authority could act as a surrogate for BNOC. It is difficult, however, for the layman to judge the extent of its knowledge.

There has been little discussion in the United Kingdom

about whether or not there should be a depletion policy. Government policy until the end of 1979 was that exploration and development should proceed as rapidly as possible, with the aim of achieving self-sufficiency at an early date, say, by late 1980 or early in 1981, but there was little indication as to what government policy on depletion would be thereafter. Table 25 and earlier discussion showed how oil production is expected to build up rapidly from 1980 in excess of home demand. But from a peak of between 0.5 and 1.0 mbd in excess of domestic demand in the mid-1980s, the United Kingdom could become a net importer of oil before the end of the 1980s unless substantial new finds are made and developed.[1] Thus the United Kingdom could become dependent on imports by the late 1980s at a time when oil supplies from OPEC and elsewhere may be neither secure nor sufficient. Such a dependence could be avoided only through a depletion policy extending through the 1980s. Such a depletion policy is, however, ruled out without effective monitoring by a state company or by the Department of Energy or by both acting in combination, or just possibly by the Atomic Energy Authority.

The issue of regulating the rate of depletion may, however, be academic. It has been seen that, ever since the discovery of North Sea oil, the emphasis has been on quick depletion. It is doubtful whether it will ever cease to be. The United Kingdom has been undergoing the same process of 'de-industrialisation' as Canada – that is, the growth of its imports of manufactured goods has persistently tended to exceed the growth of its exports. UK manufactured exports were twice the value of manufactured imports in 1965, but were only 50 per cent greater in 1970, 15 per cent greater in 1978, and 5 per cent greater in 1979 (See Chart 4). 'In simple terms this implies that not only do foreigners not want our goods, neither do we.'[2] In these circumstances oil has been used to balance or mitigate what would otherwise have been

---

1 See, *Exploration and Development of UK Continental Shelf Oil*, Energy Commission Paper No. 17, by UK Offshore Operators Association Limited, October 1978, p. A8.
2 F. Blackaby (ed.) *De-industrialisation*, Heinemann, London, 1978, p. 260.

an even more unfavourable balance of payments, and the exchange value of the pound sterling has risen way above a value which would reflect the true costs of UK manufacturing output relatively to the costs of her competitors.

Chart 4: *United Kingdom: Value of manufactured exports as a percentage of value of manufactured imports.* (SITC groups 5, 6, 7, 8)

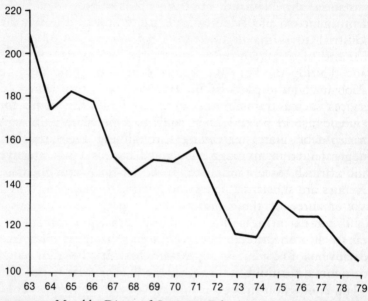

SOURCES: *Monthly Digest of Statistics,* February 1980 pp. 118, 119; *Monthly Review of External Trade Statistics,* No. 15, December 1976, pp. B8, E2.

Is the high exchange value of the pound sterling, resulting in part from oil, in part from the attempt to control inflation through high interest rates, likely to aggravate the process of 'de-industrialisation'? Both Norway, exporting oil, and Holland, exporting North Sea gas, were initially in a better position than the UK. Both, however, have experienced a serious deterioration in the non-oil balance of payments. '. . .

The United Kingdom: the Odd Man Out

some economists have traced [their difficulties] to their North Sea riches.'[1] These precedents do not make the outlook for Britain propitious.

Could North Sea oil tax revenues, assessed in Table 26, be used to remedy the manufacturing disparity, on the assumption that it is not already too late? An official paper[2] has suggested four uses for oil revenues. First, investment in industry through general investment incentives, selective assistance, the National Enterprise Board and Development agencies, and regional policy. Secondly, improving industrial performance by reducing personal taxation to guarantee real take-home pay without inflationary wage rises. Thirdly, investing in energy, covering both conservation and additional supplies. Fourthly, increasing 'essential services' such as training programmes and the social services. It would also be possible to spend the revenues on repaying foreign debt, increasing overseas investment, reducing the budget deficit, or investing in state industries such as British Rail, British Steel and British Leyland. Although the revenues are substantial, they are sufficiently large for only two or three of these options to be undertaken to any significant extent. It is thus necessary for the Government to make a choice. The better policies to pursue depend on what conclusions one draws about the reasons for the UK's economic difficulties.

The conclusion drawn in this book is unequivocal. The difference between a developed country and an undeveloped country is the possession by the former of a superior technology, and if the poverty of the underdeveloped country is to be removed what is required is an equalisation of technology, an equalisation which the developed are reluctant to contemplate. Similarly within the developed countries a better performance by one country than another is due to a superior technology. The poverty of British performance is due fundamentally to an inferior technology – not necessarily

1 Ibid., p. 214.
2 HMSO, *The Challenge of North Sea Oil*, Cmnd. 7143, March 1978.

to a lack of inventiveness, but to a failure to apply that inventiveness to manufacture and marketing.

The Conservative Government elected in Britain in 1979 appears to believe in the resurrection of the days of the Industrial Revolution, when individuals were free to do as they wished, to hold down wages as they wanted, and to re-invest the surplus as they pleased. In none of the countries examined in this book is this precedent followed. In all of them the individual has been reluctant to invest for the distant future and the State has perforce had to fill the breach. Why should the Britishers of today be more like their forbears than like their contemporaries in other countries? On the face of it, one can see no reason. On the contrary, there has been a development which likens them to their contemporaries — namely, the increasing importance of technology. For technology requires time. It requires a long look ahead — twenty years — and the patience to wait for what may be only a possible return. It is with the help of the State and through Governmental institutions that an investment of such long maturity can most easily be embarked upon.

It is to be noted that, while other countries — Canada, Kuwait, Venezuela, the United Arab Emirates — have segregated at least part of their oil revenues in a special fund, the United Kingdom has not. Oil revenues have been treated in the same way as revenues from taxes on income or sales; they have been merged in the general pool of money at the disposal of the Government. Yet there is a fundamental difference between oil revenues and other kinds of revenues, in that they are revenues derived from the sale of a capital asset. They should therefore, at least in part, be converted into a different kind of capital asset.

This conversion requires appropriate institutions. In particular, it requires a form of planning. The conventional approach to planning is to assume different rates of growth in the GNP and from these rates to derive other magnitudes. This approach leaves out of account the dimensions of politics, society and technology. It implies, for example, that if the rate of growth in GNP is 5 per cent a year, a straight extrapolation will give the magnitudes relevant to a rate of growth in GNP of 10 per cent a year. There is thus left out of account the effects on society of different rates of growth and the changes in society which may affect growth rates.

An ideal approach to planning should start from the political; it should present scenarios according to different political developments. From these scenarios can be derived a combination of economic and societal consequences, the economic acting on society and vice-versa. Finally, possible technological changes foreseen should be fed into the scenarios.

From scenarios drawn up in this way by an independent futures unit within the Government, the Cabinet should select the scenario which it deems most appropriate. An independent technological unit should then indicate the technologies required to meet the needs indicated by the chosen scenario. These technologies would be financed from the capital fund of North Sea oil revenues, or at least part of these revenues segregated into a separate and identifiable fund.

UK planning is far from approaching this ideal model, which would involve drawing up qualitative scenarios of the world twenty or more years ahead, as is done by such companies as Shell and ICI, and an independent scientific Ministry which aims at meeting the needs emerging from the scenarios. A minor analogue to such a Ministry exists in Canada; its function is to analyse and sift the scientific programmes of different parts of government so that a total and coherent scientific policy can be laid before the Cabinet. No comprehensive futures unit and no independent scientific Ministry exists in the United Kingdom. The country's ability to transmute oil into a different form of capital asset is thus weakened.

In the Budget of March 1980 the Chancellor of the Exchequer said of UK policy 'we shall take the opportunity offered by the growth of oil revenues to bring the level of public sector borrowing steadily down, and that is what our medium term strategy envisages.'[1] At the same time it was

1 *The Times*, 27 March 1980, verbatim report of Budget speech.

made clear that from 1979/1980 to 1983/1984 expenditure on industry, energy, trade, and employment was planned to fall by £1,200 million in real terms – that is, by 41 per cent. Yet none of these policies is likely to make any contribution to raising the level of investment in the UK and redressing her technological inferiority.

The lessons to be learnt by the United Kingdom from the experiences of OPEC members seem clear and unequivocal. Without a state-owned company endowed with appropriate powers the United Kingdom will not be able to judge the right rate of taxation and will therefore not be able to reap the full economic rent; will not be able to pioneer marginal fields; and may not be able to regulate the rate of depletion. Nor, without new institutions, will she be able to retain through investment the capital that is represented by oil. She thus threatens to lose out on all counts and so miss the opportunity which oil has presented to her.

The United Kingdom

Arnold, Guy, *Britain's Oil*, Hamish Hamilton, London, 1978.

Blackaby, Frank, (ed.), *De-industrialisation*, Heinemann, London, 1978.

British National Oil Corporation, *Report and Accounts*, 1979.

Dam, K. W., *Oil Resources: Who Gets What How?*, University of Chicago Press, London, 1976.

Energy Commission Paper No. 17, *Exploration and Development of UK Continental Shelf Oil*, 1978.

Hamilton, Adrian, *North Sea Impact. Off-shore Oil and the British Economy*, International Institute for Economic Research, London, 1978.

HMSO, *Development of the oil and gas resources of the United Kingdom*, 1979.

HMSO, House of Commons Parliamentary Debate, *Weekly Hansard*.

HMSO, *The Challenge of North Sea Oil*, March 1978.

HMSO, *UK Offshore Oil and Gas Policy*, July 1974.

Johnson, Christopher, *North Sea Energy Wealth 1965–85*, Volumes I and II, Financial Times, London, 1978.

Johnson, Christopher, *The Improvement of the North Sea Tax System*, Institute of Fiscal Studies Conference, June 5, 1979.

Page, S. A. B., *National Institute Economic Review*, No. 82,

November 1977, 'The value and distribution of the benefits of North Sea oil and gas 1970–1985'.

Robinson, C., and Morgan, J., *North Sea Oil in the Future. Economic Analysis and Government Policy*, Macmillan, 1978.

Robinson, C., and Rowland, C., *Petroleum Economist*, December 1978, 'Marginal effect of PRT changes'.

CHAPTER 11

The Nature of OPEC

Three forces brought OPEC into being. The first was the shift in the balance of production from the United States to the Middle East, with the result that the practice of basing the price of oil on the Mexican Gulf lost its rationale. The second was the entry in the late 1950s into the Middle East of independent producers, their incursion weakening the ability of the major vertically integrated producers (the so-called 'seven sisters') to maintain prices and hence the revenues of the Governments to which, in exchange for their concessions, they paid royalties and taxes. OPEC was created in 1960 to steady the faltering market. The third was a general tendency on the part of raw material producers in the developing world to get together.

But OPEC was not created without much groundwork. The 1950s had seen a growing experience and understanding of petroleum affairs in the oil-producing countries of the third world. Earlier chapters have described how profit sharing was introduced in the 1950s, country by country, following the lead given by Venezuela. There was also greater cooperation and cordination of policies in this period, much of it under the auspices of the Arab League, although its members were always aware of the crucial importance of the two non-Arab States: Iran and Venezuela. At a meeting held in Baghdad from September 10 to 14, 1960, OPEC was born, the original members being Iraq, Iran, Kuwait, Saudi Arabia and Venezuela. Additional members followed: Qatar in 1961, Libya and Indonesia in 1962, Abu Dhabi in 1967, Nigeria in 1971, Ecuador in 1973 and Gabon in 1975.

The objectives of OPEC were set out in their Statute, and were essentially threefold: the coordination and unification of

The Nature of OPEC

the petroleum policies of member countries; the stabilisation of prices in international oil markets 'with a view to eliminating harmful and unnecessary fluctuations' and the maintenance of a steady income to the producing countries and an efficient, economic and regular supply of petroleum to consuming nations.[1]

It is clear that OPEC's initial purpose was to prevent posted (or published) prices as determined by the companies, and on which the tax revenues of the governments were based, from falling further. This purpose was attained, for while the posted price of Saudi Arabian light fell from $2.08 per barrel in June 1957 to $1,80 in August 1960, it never fell below that level, despite the fact that a weak demand for crude oil in the 1960s caused the market price, in contrast to the posted price, to decline continuously throughout that decade. Collective action was thus paying off. By the close of the 1960s OPEC was giving greater consideration to raising prices. In the OPEC Policy Statement of June 1968 'the Conference recommended that posted and tax-reference prices should be fixed by the governments, and that these should be linked to a prices index of manufactured goods traded internationally.'[2] From 1970 OPEC began to secure some increases in the posted prices of crude oils, and indeed the period from 1970 to 1973 is of considerable importance in OPEC's development.

Three events occurred in 1969/1970 which shifted the balance of power from oil-consuming to oil-producing countries, and enhanced the power of OPEC. First, the Libyan Revolution of September 1969 brought to power a government which took a much stronger line in negotiations with the oil companies. Faced with continued intransigence on the part of the companies over an increase in price, Libya cut back production, beginning in July 1970; these cutbacks had reached 1.0 mbd by the end of the year. Second, in May 1970 the Trans-Arabian Pipeline, which carried 0.5 mbd of Saudi

---

1 OPEC, *The Statute of the Organisation of the Petroleum Exporting Countries*, February 1978, p. 3.
2 Mana Saeed Al-Ataiba, *OPEC and the Petroleum Industry*, Croom Helm, London 1976, p. 117.

Arabian crude oil to the Mediterranean, was accidentally cut in Syria. The Syrian Government would not permit repairs, using it as an excuse to raise its transit revenues. Supplies were thus further reduced. Third, the demand for oil in Western Europe was high in 1970. The combination of reduced supply and increased demand enabled Libya to raise the posted price of its crude by more than 30 cents a barrel before the end of 1970. Other OPEC members followed.

At their Conference in Caracas in December 1970 OPEC members realised the need for coordinated action to take advantage of existing market conditions. Accordingly, they resolved to secure an increase in posted prices, the elimination of discounts off posted prices, and an increase in the tax rate from 50 to 55 per cent. Negotiations on these issues with the oil companies opened in Tehran on 12 January 1971. OPEC faced a united front of company resistance to its proposals, and it was only in the face of threats of unilateral legislative action that the companies finally gave in. The resulting Tehran Agreement was to cover the five years 1971 to 1975, and allowed for an immediate increase of all Arabian (Persian) Gulf oil prices by 33 cents a barrel, with further annual increases of 5 cents a barrel; the tax rate was to be raised to 55 per cent, and other changes were made. The 1971 agreement was the 'first serious OPEC attempt at collective bargaining.'[1] Its main significance was the fact that henceforth prices would be determined, not by the companies, but by governments.

Even so OPEC did not reveal its full power until 1973–74, when prices were virtually quintupled. Was this steep increase in prices the action of a cartel? In part 'yes', since collective action by governments had wrested from the companies the determination of prices. But there were also other factors at work.

Formally OPEC meets to decide on the price of one kind of

---

1 Joe Stork, *Middle East Oil and the Energy Crisis*, Monthly Review Press, 1975, p. 173.

oil only – light Arabian crude.[1] The price of Arabian light is intended to be a point of reference in relation to which the prices of other oils can be determined. Thus, lighter crudes and crudes with a lower sulphur content can command a higher price than Arabian crude, as can crudes which are nearer the consumer, such as those of Libya and Algeria; conversely, the prices of those crudes that are heavier will have a lower price. Table 27 lists the prices of crude petroleum for various OPEC countries from 1970 to 1979. Up to 1976 the prices given are posted prices; from 1977 they are official state selling prices. The crudes from each country are selected so as to represent the wide variation in prices which are charged at any given time. It can be seen that the crudes lighter than Arabian crude, such as those of Libya and Venezuela, have a higher price than the heavier crudes of Iran and Kuwait. The standard Saudi Arabian crude, Arabian Light, has a price which broadly lies in between these extremes. Normally, it is only the price of Arabian Light which remains fixed until the next meeting, while the prices of all crudes other than Arabian crude may vary between meetings, in the light of the market, without determination by OPEC. During 1980, however, Saudi Arabia began to change the price of her crude between OPEC meetings.

What then is the market? Geographically it is principally the Rotterdam 'spot' market, Rotterdam being a major centre of oil processing, refining and storage. It exercises an influence on prices out of all proportion to the volume of oil which is transacted through it. Some 15 per cent of OPEC production is sold by government to government, some 60 per cent is sold to the major producing companies with which the OPEC governments have contracts, and the rest is sold to independents, some of it on the 'spot' market. The major companies endeavour to maintain a balance between their supply of crude oil and the various derivatives which they wish to

---

[1] An oil is labelled light or not according to its density. Arabian light crude has traditionally been defined as crude containing 34° API (American Petroleum Institute). This definition is now falling out of fashion.

Table 27: Prices of crude petroleum for various OPEC countries

Monthly averages in US dollars per barrel, current prices (a)

	Libya Zuetina (40°)	Saudi Arabia Arabian Light (34°)	Venezuela Oficina (35°)	Iran Medium (32°)	Kuwait (31°)	Iraq (35°)	Nigeria Bonny (34°)	Indonesia (API various)
1970	2.23	1.80	2.80	1.63	1.60	1.74	2.25	1.70
1971	3.23	2.19	2.80	2.13	2.09	2.16	3.05	2.09
1972	3.62	2.47	3.22	2.41	2.36	2.44	3.39	2.78
1973	5.15	3.27	4.45	3.22	3.14	3.24	4.80	5.87
1974	15.77	11.58	11.22	11.56	11.47	11.60	14.69	11.70
1975	15.32	11.53	14.50	11.51	11.39	11.55	12.17	12.65
1976	16.20	12.38	15.63	12.22	12.11	12.39	13.81	12.80
1977	14.00	12.39	13.99	12.49	12.34	12.64	14.56	13.55
1978	13.90	12.70	13.99	12.49	12.26	12.60	14.17	13.55
1979	21.28	16.97	19.66	18.53	18.17	18.35	21.26	18.59

(a) f.o.b. export prices; beginning in 1977 state sales prices, other than Indonesia.
SOURCES: *United Nations Monthly Bulletin of Statistics*, April 1979, p. 171, December 1977, pp. 171, 172, January 1980, pp. 171, 176; *Middle East Economic Survey*, Vol. XXIII, No. 11, 31 December 1979, pp. 6, 7, 8.

extract from it. An exact balance, however, is difficult to achieve. If a company has a surplus of crude in relation to the derivatives likely to be required of it, it sells on the Rotterdam market. Conversely, if it is short of crude in relation to its needs downstream, it buys on the 'spot' market. The greater part of the oil produced passes down the vertical chain of the major companies, but a residue appears on the spot market, and the price struck there is in turn reflected in the price agreed upon for longer-term contracts between the governments and their major contracting companies. If, for example, there is a shortage in the market, those crudes which are lighter than Arabian crude will enjoy an enhanced premium, while those crudes which are heavier will have a lower discount from the price of Arabian crude. The reference or 'marker' price will then be out of line with actual prices, and there will be nothing for it but to adjust upwards the price of Arabian crude at the next OPEC meeting.

This is what happened in 1973–74. Commodity prices, including the market price of oil, had been rising since the early 1970s. From 1971 to January 1973 there were increases in the 'spot' market price of crude oil in excess of those provided for in the Tehran Agreement. A revision of the Agreement seemed indicated, and to this end negotiations were entered into in 1973. Shortly after the outbreak of the Arab/Israeli war of October 1973 the oil companies withdrew from the talks, knowing that there would be a price increase anyway. The final decision to raise prices was thus a unilateral one on the part of OPEC, but it was a decision which followed the market and did not intiate the market movement.

The same thing happened in 1979. A number of influences combined to cause an increase in the market price: the failure of the consuming countries to take effective action on conservation in the use of oil; the harshness of the winter of 1978–79 both in Europe and in the United States; and, perhaps most important of all, the decline in Iranian production, offset to an unknown extent by increases in output on the part of the other OPEC countries. In raising the 'market' price

OPEC was once again following the 'spot' market, which, it will be seen from Table 28, rose from 12.98 US dollars a barrel in October 1978 to 35.40 dollars a barrel in June 1979.

In between the two increases of the 'marker' price of 1973–74 and 1979, the real price – that is, the nominal price reduced by a given rate of inflation – fell. It will be seen from Table 3 that between January 1974 and January 1978 the real posted price fell by around two US dollars a barrel. What rate of inflation should one use in translating actual prices into real prices? Not the internal rates of inflation in Western countries, nor the rate of inflation in products entering into trade between industrialised countries, but the rate of inflation shown by the goods being imported by oil-producing countries – mainly capital equipment and chemicals. The rate of inflation in this last significant sense has generally moved faster than any other rate of inflation. Unfortunately there is no commonly accepted measure of it. In default of such a measure use has been made in Table 29 of the unit value index of manufactured goods for industrialised (OECD) countries published by the Statistical Office of the United Nations since 1948.[1] As indicated, this measure understates the fall in the real price of the OPEC 'marker' crude. Even so, the fall is undeniable.

If it be a true interpretation of events that OPEC has in general followed the market upward rather than led it, OPEC can only with qualification be described as a cartel. Yet its members attach importance to its continued cohesion. One reason for this is clearly the fear of a collapse, if the market were falling, with one country undercutting another. It is significant that the most serious occasion when a difference of view appeared between the several members of OPEC was at a meeting in Caracas in December 1979, when prices on the so-called spot market had been rising. The members decided each to go its own way, according to its expectation

---

[1] See G. F. Ray, 'The 'Real' Price of crude oil', *National Institute Economic Review*, No. 82, November 1977, p. 59; and UN *Monthly Bulletin of Statistics*, March 1975, 1976, 1977, 1978, 1979, December 1979, special tables. See also, *Middle East Economic Survey*, Supplement to Volume XXI, No. 49, 25th September 1978.

Table 28: *Crude oil 'spot' price*

	Spot price, Saudi Arabian Light 34° crude; weekly average, US dollars per barrel
October 1978	12.98
November 1978	14.43
December 1978	14.83
January 1979	16.24
February 1979	22.56
March 1979	22.38
April 1979	21.25
May 1979	28.94
June 1979	35.40

SOURCE: *OPEC Review*, Vol. III, No. 2, Summer 1979, Part II.

of what the market could bear. The resulting differences in prices and the desire of Saudi Arabia in particular to bring them into line again precipitated a greater frequency of meetings in 1980. If demand continues to outstrip supply, including the maximum which Saudi Arabia is prepared to produce, then the upward pressure on prices will continue, and Saudi Arabia will be powerless to prevent it in spite of her desire for greater price stability.

The future, so far from seeing a loosening of OPEC, seems likely rather to see an extension of OPEC activities and possibly a greater degree of cohesion on the part of its constituent members. The main reason for this view is that OPEC, while nominally concerned with prices, is in reality concerned with wider problems – the internal development of its member states, the development of alternative sources of energy so that the life of oil – the main natural asset of OPEC countries – can be prolonged, and the development of those members of the Third World which are without oil.

The internal development of OPEC countries requires primarily, though not exclusively, the maintenance of the governments' oil revenues. These have been eroded both by inflation and by changing foreign exchange rates. The pro-

Table 29: *Real price of Saudi Arabian Light crude oil in US dollars per barrel (a)*

	Posted price		State sales price	
	Current prices	January 1972 prices	Current prices	January 1972 prices
9 August 1960	1.80			
October 1970	1.80			
14 February 1971	1.80			
15 February 1971	2.180			
1 June 1971	2.285			
20 January 1972	2.479	2.479		
1 January 1973	2.591	2.422		
16 October 1973	5.119	4.031		
1 January 1974	11.651	9.032		
1 November 1974	11.251	7.212	10.463	6.705
1 October 1975	12.376	7.883	11.510	7.331
1 January 1977	13.000	7.471	12.090	6.948
1 July 1977	13.660	7.631	12.700	7.095
1 January 1978	13.660	6.969	12.700	6.480
December 1978			12.700	5.799
1 January 1979			13.339	6.008
1 April 1979			14.546	6.552
1 June 1979			18.000	7.759
1 November 1979			24.000	10.133
1 April 1980			26.000	
May 1980			28.000	

(a) The current prices have been deflated by a unit value index of manufactured goods exports for industrialised (OECD) countries. For an explanation of the use of this index, see text.

SOURCES: OPEC, *Annual statistical Bulletin 1977*, pp, 129, 130, 131; *Middle East Economic Survey*, Vol. XXIII, No. 11, December 1979, p. 6; *The Times*, 29 January 1980; *National Institute Economic Review*, Number 91, February 1980, p. 124.

tection of these revenues, insofar as they are invested temporarily or permanently abroad, is an important reason for continued cohesion; for only if the OPEC countries are united can they make such representations as may seem appropriate.

The OPEC countries have been able to acquire substantial

Table 30: *OPEC members' surpluses in nominal and real terms 1973–79*

Total OPEC	Thousand million US dollars								
	1973	1974	1975	1976	1977	1978	1979 1st half	1979 (a)	1979 (b)
Current prices:									
Oil revenues	23	95	97	113	129	124	67	173	197
Total exports(c)	42	125	116	140	153	149	84	207	214
Current account surplus	1	61	38	40	33	12	16	45	62
1975 prices (d):									
Oil revenues	32	106	97	112	118	99	49	126	na
Total exports(c)	58	140	116	139	140	119	61	150	na
Current account surplus	1	68	38	40	30	10	12	33	na

na: not available.
(a) Projected by NIESR August 1979.
(b) Estimated by Morgan Guaranty Trust, in *World Financial Markets,* March 1980.
(c) Includes oil revenues, other non-oil exports of goods and exports of services.
(d) Figures deflated by the unit value index for exports of manufacturers in the year in question.
SOURCES: *National Institute Economic Review*, Number 89, August 1979, Table 2, p. 25; *Bank of England Quarterly Bulletin*, Volume 15, No. 4, December 1975, pp. 342, 343; Volume 17, No. 4, December 1977, p. 442; Volume 19, No. 3, September 1979, pp. 276, 277; Volume 19, No. 4, December 1979, p. 361; *Petroleum Economist*, March 1975, p.85; *Middle East Economic Survey*, Volume XXIII, Number 25, 7 April 1980, p. viii.

foreign assets because the revenue they have earned from oil has been greater than their expenditures on imports from the oil consuming countries. Table 30 shows how rapidly OPEC members' oil revenues have increased since 1973. The price

Table 31 Deployment of OPEC members' surpluses by country and type of investment

	Thousand million US dollars							
	1974	1975	1976	1977	1978 (a)	1979(a) first half	1979(a) whole year	Total at end 1978
Country:								
UK	21.0	4.3	4.5	4.3	−1.8	4.3	17.3	32.3
USA	11.6	10.0	12.0	8.9	1.3	−1.1	7.7	43.8
Other countries	20.9	17.4	17.3	19.7	13.6	7.5	19.1(d)	88.9
International organisations	3.5	4.0	2.0	0.3	0.1	−0.3	−0.5	9.9
Type of Investment:								
Bank deposits	28.5	9.9	12.8	9.4	3.9	4.4	na	64.5
Government bonds	9.1	2.0	2.3	3.5	−2.6	−0.7	na	14.3
Other investments (b)	19.4	23.8	20.7	20.3	11.9	6.7	na	96.1
of which aid (c)	5.9	8.2	8.1	7.6	na	na	na	na
Total	57.0	35.7	35.8	33.2	13.2	10.4	na	174.9
Cumulative total	57.0	92.7	128.5	161.7	174.9	185.3	na	174.9

na: not available
(a) 1978 and 1979 data from Bank of England are not exactly the same as OAPEC data, but the differences are not significant.
(b) Includes holdings of equities, property, etc., and loans to less developed countries.
(c) OECD data; concessional and non-concessional aid.
(d) First three quarters only.
SOURCES: OAPEC, *Fifth Annual Report*, 1978, Table 18, p. 67; *Bank of England Quarterly Bulletin*, Volume 19, No. 3, September 1979, pp. 276, 277; Volume 19, No. 4, December 1979, pp. 361, 389; Volume 20, No. 1, March 1980; I. F. Shihata and R. Mabro, 'The OPEC Aid Record', *World Development*, Volume 7, p. 172.

rises in 1979 and 1980 meant that these revenues would be substantially greater in 1979 and 1980 than they were in 1978. During the same period imports into OPEC countries similarly increased substantially, but not sufficiently to prevent current account surpluses in every year. Table 31 shows how the surpluses have been distributed in investments in foreign countries. A comparison of the two Tables shows that the total invested each year is almost identical to, or a little less than, the current account surplus for that year, as would be expected.

It is significant that nearly half of the surplus funds have been invested in the United States and the United Kingdom. In 1974 the United Kingdom was the most favoured country for OPEC funds, but since then the United States has clearly been the major recipient. International organisations such as the IMF and the World Bank received payments in the first few years so as to establish the OPEC members' position with these organisations. Subsequently, however, such contributions have become negligible. The deployment of surpluses in 'other countries' refers in large part to aid to other Arab and non-Arab developing countries. In the latter part of the 1970s OPEC members tried to diversify their holdings away from the US and the UK, towards countries such as West Germany, Switzerland and Japan. In 1978 there was a net withdrawal of funds from the UK, and an investment in the US only one-tenth of that in other countries. 1979 saw a net withdrawal from the US in the first six months.

In its 1978 *Annual Report* OAPEC described the position as follows: 'Prior to 1974, the bulk of OPEC investments was concentrated in the United Kingdom. Presently, the United States is the leading centre, with OPEC investments at end-1977 reaching $43 billion, as compared with about $34 billion in the United Kingdom. [This had changed little by end-1978, as Table 31 shows.] About half of OPEC foreign assets are placed in these two countries, while the other half is distributed throughout the other Western industrial countries and a number of international institutions.'[1] This concentration in two countries clearly provides a threat to OPEC control over their own assets. Well in advance of the US action in freezing Iranian assets in 1979, OAPEC had voiced fears. 'Arab investments are exposed to the possibility that local measures might be taken to freeze their movement, decrease their returns, or cause them great losses from exchange rate

---

1 OAPEC, *Secretary General's Fifth Annual Report 1398H–1978AD*, Kuwait, May 1979, p. 63.

fluctuations.'[1] It is this fear which has caused OPEC members to limit investment in the US and UK, and look elsewhere. But the restricted nature of opportunities for the investment of large sums of money, combined with the fact that investments in other countries are similarly open to interference, influences OPEC members towards restricting their financial surpluses by restricting oil production.

A second element in OPEC's members' deployment of their surpluses has been the type of investment which they have favoured or into which they have been forced. It is clear from Table 31 that most of the surpluses have been invested in bank deposits and Government bonds. After the removal of aid to other developing countries from 'other investments', well over half of OPEC's surplus has been deployed in short-term assets. To an extent this investment in short-term assets is inevitable, for the funds have to be kept liquid pending more permanent investment, and permanent investment overseas is made difficult by the reluctance of recipient countries to permit OPEC members to invest substantially in equities or other direct investment. In the early years, when the financial surpluses suddenly emerged, they surprised both the oil-producers and the oil-consumers with sophisticated financial markets. Neither the oil producers nor the oil consumers were ready or prepared to place large sums of money other than on a short term basis in easily accessible and transferable assets such as bank deposits. However, as early as August 1974 a leading Arab financial expert was writing of the desire of Arab countries to invest their funds in assets of longer maturity. They wished to place substantial funds in direct investments. He argued that 'as yet their economies (OECD countries) have not shown signs of willingness to accept such types of investment.'[2]

The third element in the surpluses is their distribution by oil-producing country. The net external assets of OPEC countries are concentrated in the hands of four countries: Saudi

1 OAPEC, *Secretary General's Fourth Annual Report,* 1977, Kuwait, 1978, p. 76.
2 Hikmat Sh. Nashashibi, 'Other Ways to Recycle Oil Surpluses', *Euromoney,* August 1974.

The Nature of OPEC

Arabia, Kuwait, the United Arab Emirates and Iran. At the end of 1979 their net foreign assets were, respectively, $75 billion, $40 billion, $13 billion and $16 billion.[1] The first three of these countries have relatively small populations and all receive greatly more revenue from oil production than they can easily spend domestically on development and defence. Iran differed at that stage in having a much larger population and a greater capacity to use revenues internally. Other OPEC countries either had much smaller surpluses, or were net foreign borrowers.

The major surplus countries have been very cautious and conservative in their use of oil revenues. They have neither tried to upset the financial markets of the world, nor purchased large shareholdings in the major companies of industrialised countries. However, these foreign investments now produce a substantial income for the OPEC members, the few countries mentioned above earning $14.5 billion in 1979,[2] more than they earned from oil before 1974. Consequently, they must be expected to give considerable importance to opportunities for foreign investment.

Table 30 shows that in terms of 1975 prices the value of OPEC's current account surplus fell steadily and substantially from 1974 to 1978. The projected value of the surplus in 1979 was less than half the 1974 surplus in constant prices. On the basis of the UN index of OECD manufactured exports, the accumulated surplus of $185.3 billion at the end of the first six months of 1979 would have been worth $137.3 billion in 1975 prices. OAPEC claims that 'the estimated $131 billion in foreign assets held by the Arab countries at the end of 1977 only equalled, in terms of purchasing power, $90 billion of end-1974 dollars calculated on the basis of inflation in the OECD countries.'[3]

In addition, there has been a depreciation of the dollar which has affected the value of OPEC holdings. Over half of

---

1 Data from *Middle East Economic Survey*, Volume XXII, Number 28, 28 April 1980, p. (1).
2 Ibid, p. (ii).
3 OAPEC, *Secretary General's Fifth Annual Report 1398H–1978AD*, p. 64.

all OPEC foreign assets are denominated in US dollars, for in addition to assets in the USA many of the investments in Europe are in the Eurodollar market. Between 1974 and 1978 the dollar depreciated against other currencies by about 12 per cent. This further reduced the value of OPEC's assets. Various sources suggest that OPEC assets in this period earned a nominal return of between 7 and 8 per cent, and after allowing for inflation and depreciation of the dollar there was a real net negative return of between 2 and 4 per cent. 'OPEC's dollar assets earned on average a yield of 7.5 per cent during the period in question. After adjustment for interest earned, the depreciation of (the capital value of) OPEC's dollar-denominated assets has averaged 4.5 per cent annually.'[1]

To produce so much oil that surplus revenues are acquired which yield a negative real rate of return cannot in the long run conduce to the development which is the primary aim of most of the OPEC countries. From the point of view of development the ideal form of investment for an OPEC country overseas is an equity holding in an industrial firm or even a loan to such a firm. In exchange for its capital thus increased the firm should be prepared to transfer technology to the oil-producing country. It is true that Arab governments have investments of about US $40 billion in private firms. There is also substantial investment in property, particularly in the United Kingdom. By and large, however, firms in the industrialised countries are reluctant to see OPEC shareholdings, and OPEC countries have been forced, either through legal restrictions or popular pressure, to limit their holding of shares. The corollary is two-fold: the enforced holding of short-term assets, such as time deposits or Treasury bills, yielding a low rate of return; and a reduction in oil output, with adverse consequences for the industrialised world both in terms of price and quantity.

A reduction in oil output is in any case likely because of an increased consciousness of the need to conserve oil in the

---

1 Odeh Aburdene, 'The impact of inflation and currency fluctuations on OPEC's dollar assets during the period 1974-1978', *MEES*, Vol. XXII, No. 6, 27 November 1978, Supplement, p. 3.

The Nature of OPEC

ground, as will be discussed in the next chapter. The need for a policy on depletion is now being increasingly discussed at OPEC meetings. It is clear that a major part of the Algiers meeting in June 1980 was concerned with the question of future production, particularly of the two largest producers, Saudi Arabia and Iraq. One difficulty in the way of controlling production is the absence in almost all OPEC governments of any monitoring authority on the lines, say, of the Texas Railroad Commission. Even when oil has been nationalised, a nationalised oil company cannot be trusted to regulate the rate of extraction; technical ambition may prompt it rather to expand production. Its activities need, therefore, to be audited or monitored by government. The exercise by government of a monitoring activity requires, however, an additional corps of experts, and experts do not abound.

The adoption of policies to regulate the rate of depletion might lead, at any rate initially, to a reduction in the amount of OPEC crude oil available to consuming countries. In that event either each OPEC country might have to resort to rationing its foreign customers, or, more probably, the central OPEC authority would assume charge of the rationing policy. A similar development seems possible in the area of products derived from oil. So far refineries and petro-chemical plants have been set up without any co-ordination between constituent OPEC or OAPEC members, OAPEC having confined itself to the giving of information. In this sense OPEC has been less tight as a cartel than were the 'seven sisters'. When, however, the refineries and petro-chemical plants come on stream, operating in competition with plants already established in Western countries, there may well appear an excess of capacity. In that case the central authority, OPEC or OAPEC, would need to introduce a scheme of production quotas. These possibilities suggest that, so far from disintegrating, OPEC may well evolve into a tighter structure.

A further reason why the OPEC countries are likely to stick together is that they are both the self-appointed and the acknowledged leaders of the Third World. Table 31, based on OECD data, shows that OPEC aid increased very substantially after the 1973–74 price rise (it was a mere $443.5 million in 1970), and now forms a major part of all aid. Total aid from OECD countries in 1977 was $18,015 million, compared with $7,588 million from OPEC.[1] In 1976 Saudi Arabia was second only to the United States in the provision of bilateral aid, while OPEC members turned out to be the top six countries which gave the highest proportion of aid in relation to their GNP. It has been claimed that the OPEC aid has not offset the damage done to developing countries by the oil price rises of the 1970s. This seems most unlikely, given the volume of aid, and the large share of budget or balance of payments aid given by OPEC, as opposed to project aid. In addition, aid is a greater burden to OPEC than to OECD, in that it stems from oil revenues gained from the depletion of what is essentially a capital stock, rather than from a recurring flow of income.

The provision of aid by OPEC members serves a political rather than an economic purpose: to strengthen and unify the Arab world and to increase the influence of the donor countries in their dealings in world councils with the industrialised parts of the world. This solidarity could be particularly important in discussions on the future of world currencies.

Should the West be alarmed at OPEC solidarity? Only in so far as it reflects such problems as a scarcity of oil and an excess of capacity to produce derivatives from oil. What then should the West's reaction be? Not to set up a countervailing institution, as Mr Kissinger did after 1973–74; for there is no countervailing power that could be exercised. The reaction should rather be to seek co-operation with a reinforced OPEC in the solution of certain common problems.

Those problems are three. First, 'gas-guzzling' – the inordinate increase in the use of oil, which threatens to impair OPEC's ability to export, is quite as bad in Venezuela as it is in the United States. Second, a possible over-capacity to pro-

---

1 I. F. Shihata and R. Mabro, 'The OPEC Aid Record', *World Development*, Vol. 7, p. 164.

duce derivatives from oil – the OPEC countries enjoying the advantage of proximity to the source of the raw material, together with a low price, while the industrialised countries enjoy the advantage of proximity to the final market. Third, the problem of developing sources of energy in the non-oil-producing countries of the Third World. In 1977 the OPEC countries gave to the non-oil-producers of the Third World almost one-third as much as the OECD countries. There is, however, no indication that OPEC efforts are co-ordinated except perhaps to the Arab World through the Arab Development Fund. Nor is there any co-ordination with Western countries.

The World Bank is now concerned to find oil or any other energy source in the Third World outside OPEC. Yet this is an area in which OPEC countries now possess a certain expertise. What is indicated, therefore, is a joint effort by OPEC and OECD to find energy elsewhere, partly to relieve the strain on OPEC supplies, partly to meet the needs of the West. For this joint purpose the West would need an organisation to combine with OPEC. The most appropriate organisation would appear to be the OECD, the IEA being ruled out of court.

To sum up: the West has wrongly seen, and wrongly continues to see, OPEC as a cartel, fixing the price; it has ignored, and continues to ignore, OPEC's contribution to, and its potential for, development. Both OPEC and the West should recognise that there is increasingly little either can do about the price, this being primarily a reflection of demand in the face of feared scarcity. Once, however, OPEC is seen as an agent of development, the West should seek to harness its help to obviate the scarcity that is feared. Correspondingly, the West should be prepared to give, and what is required from it above all is the transfer of technology, for without this there can be no development in any part of the Third World.

CHAPTER 12

Depletion Policy and the World Energy Balance

The past few years have seen the publication of many assessments of the likely balance between the supply of and the demand for energy in the period up to 2000, and the effect of this balance on the future price of energy, especially oil.[1] These studies differ from this book in several respects. They draw their conclusions from a general view whereas this book has attempted to describe the changes taking place in individual countries and to infer from these individual changes the larger outcome. As a result, the broader studies have without exception left out of account two important factors governing the supply of oil: first, the rapid increase in consumption taking place in all oil-producing countries; second, the feeling of the oil producers that they have been or are being robbed of their oil, that their industrial development as a result is being retarded, and that the life of their resources must be prolonged to allow of the changes in social habits which are necessary for successful industrialisation. In nearly all the countries studied – Saudi Arabia is the outstanding exception – restrictions on production have been introduced. The conclusion seems inevitable that the 'shortage' of oil is likely to appear earlier than has been forecast by other studies and that between now and the end of the century the price of oil relative to the prices of other things (that is, its real price) is likely to continue to rise.

1 Some of the more important studies are: Report of the Workshop on Alternative Energy Strategies, *Energy: Global Prospects 1985–2000*, McGraw-Hill, New York, 1977; OECD, *Energy Prospects to 1985,* Volumes I and II, Paris, 1974; OECD, *World Energy Outlook*, Paris, 1977; Energy Information Administration, *Annual Report to Congress*, Volume II, 1977, United States Department of Energy, Washington, 1978; World Energy Conference, *World Energy: Looking Ahead to 2020*, IPC, 1977.

The fact of a rising trend in price need not, of course, exclude the occasional glut. A glut, for example, began to appear in the second quarter of 1980. Stocks had been built up in 1979 because of the Iranian revolution and in 1980 the spread throughout the world of a recession led to excess supply and thus falling prices. A glut from similar or other causes can occur again. The trend in prices, however, is likely to be upwards.

The important elements in the world energy balance are demand, which depends on economic growth and the efficiency with which energy is used; supplies of non-oil energy such as gas, coal, nuclear power and renewable sources – matters which fall outside the scope of this book; and supplies of oil. Previous studies have estimated future levels of demand, non-oil energy supplies, and oil supplies from countries which are not members of OPEC, and have then treated demand for OPEC oil as a residual. The implication of such an approach is that any failure by OPEC to fill this 'gap' would produce an energy 'crisis'. It is quite true that oil has been used as a residual fuel to fill any gap left between the demand for and supplies of other fuels. Thus, fluctuations in the world economy leading to greater demand for electricity or fuel for industry would often be met by oil from OPEC. Table 32 shows how OPEC production built up from 1967 to 1973 in response to the growth in the world economy. The reductions in output in 1974 and more especially in 1975, and again in 1978 reflect the poor state of the world economy in those years. However, with the transfer from the companies to governments of the right to determine the level of oil production, and with the greater awareness on the part of the governments of the need to control the rate of extraction, OPEC production can no longer be regarded merely as a residual. It has become an important component of supply in its own right.

In the period immediately after the price rises of 1973–74 OPEC members concentrated on using the vastly increased revenues – sums far greater than they had ever previously had at their disposal – on industrialisation and development expenditures, purchases of arms, current consumption expenditure, and largesse in distributing aid to the Third World generally. By the end of the 1970s, however, it was realised that industrialisation and development could not be achieved by money alone. It required as complements skills, an adaptation of the educational system which – in Muslim countries – was in accord with the tenets of Islam, an infrastructure, and a strategy on the industries to be developed. This must inevitably be a long process, and its length must necessitate conservation of the oil in the ground.

Table 32: *OPEC crude oil production*

	Million barrels per day									
	1967	1970	1972	1973	1974	1975	1976	1977	1978	1979
Total OPEC production	16.8	23.6	27.1	31.0	30.7	27.2	30.7	31.4	29.9	30.7

SOURCES: *BP Statistical Review of the World Oil Industry 1977*, p. 19; OPEC, *Annual Statistical Bulletin 1977*, p. 13; *Petroleum Economist*, Volume XLVII, Number 3, March 1980, p. 135.

The increased consciousness of the need to conserve oil in the ground is not the result of a comparison between the return available from other forms of investment and the expected price of oil, the conventional doctrine discussed in Chapter One. Even the erosion in the value of foreign investments caused in part by inflation, in part by the fall in the exchange value of the dollar (the currency in which the price of oil is denominated) has not been an important factor in this change. The change has been caused fundamentally by a heightened awareness of the importance of oil as the only source of the surplus which can make industrialisation possible.

In this respect the developing countries of OPEC cannot regard conservation in the ground in the same way as does the United States which was industrialised by the time oil was discovered (Canada and the United Kingdom are somewhat different, in undergoing a process of de-industrialisation. It is only in degree that they differ from developing countries, and the required policies may be much the same).

It has been seen that the United States was the first country to adopt oil conservation practices. Its object in doing so was to maximise the technical recovery of oil by avoiding waste, and to secure the highest return to the oil companies by preventing prices from falling too far. Conservation in this sense is quite different from conservation as seen by a developing country. To the latter, if oil were depleted too rapidly then the country would face the exhaustion of export revenues and the disappearance of the major part of its national income before alternative wealth-creating industries had been established. For this reason a developing oil-producing country needs a higher ratio of reserves to production than a developed one. Whereas the industrialised countries can afford a reserves to production ratio of ten or fifteen to one, the Middle Eastern OPEC countries' existing ratios of forty to one are probably too low.[1]

Even the report *North-South* by the so-called Brandt Commission failed to grasp this point. In its recommendations it suggested that 'Oil-exporting countries, developing and industrialised, will assure levels of production.'[2] To a developing country the implication of this recommendation is that its oil is for the outside world, not for its own internal growth. The suspicion that it is being robbed is thus enhanced.

A gloss on the problem different from that of the Brandt Commission has been put by H. E. Shaikh Ali Khalifa Al-Sabah, Minister of Oil for Kuwait: 'I envisage that OPEC countries' production policies will be attuned to the requirements of their economic and social development'; 'it is not

---

[1] See, Al-Chalabi and Al-Janabi, 'Optimum Production and Pricing Policies', *The Journal of Energy and Development*, Spring 1979, p. 238; Fadhil Al-Chalabi, 'The Concept of Conservation in OPEC Member Countries', *OPEC Review*, Autumn 1979; Adnan Al-Janabi, 'Production and Depletion Policies in OPEC', *OPEC Review*, Vol. III, Number 1, March 1979; Alberto Quiros Corradi, 'Energy and the Exercise of Power', *Foreign Affairs*, Summer 1979.
[2] *North-South: a Programme for Survival*, Pan, London, 1980, p. 279.

unreasonable that an oil-producer should have an oil policy based on a ratio of reserves to production around 100:1.'[1] The crucial question, then, is how this greater awareness of the need for a depletion policy to give more time for industrialisation will affect OPEC production over the next twenty years. It is not easy to make predictions beyond the 1980s, but the evidence gathered and discussed in earlier chapters makes possible a much more detailed analysis of likely OPEC production in the late 1980s than has been produced by any of the broader-based studies on the world energy balance.

There is a tendency in the industrialised countries, more especially in the USA, to regard OPEC oil reserves as virtually unlimited and capable of sustaining substantially increased levels of production. This is mostly because of the very high level of proven reserves held by Saudi Arabia, which are much larger than the reserves of any other country.[2] However, Table 33 shows that despite these substantial Saudi reserves, the total OPEC proven oil reserves are not unlimited. The total at the end of 1978 of 445 billion barrels was enough to sustain only another forty-one years of production at the 1978 level. It is true that there could be additions to reserves in the future, both from new discoveries and enhanced recovery from existing fields, but these additions could be limited. As the Kuwaiti Minister of Oil has argued: 'we have now located the "easiest" reserves and future oil finds will prove more difficult to recover . . . because these finds will be located in smaller fields and/or harsher environments . . . and, therefore, their technical cost will be higher.'[3] Indeed, during the 1970s more oil has been depleted from OPEC than new reserves discovered. 'Between 1971 and 1978, the total net addition to reserves was in the order of 28 billion barrels,

---

[1] 'Conceptual Perspective for a Long-Range Oil Policy', published in *Middle East Economic Survey*, Volume XXII, Number 48, 17 September 1979, Supplement.
[2] Statistics on oil reserves refer to two levels of reserves, corresponding to different degrees of certainty. 'Proven reserves' refer to oil that is recoverable from known reserves with current technology and prices. 'Ultimately recoverable reserves' include new discoveries and an allocation for enhanced recovery, whether this is the result of improved technology or of a higher real price for oil which makes economic the recovery of oil which is more expensive to produce.
[3] H. E. Shaikh Ali Khalifa Al-Sabah, op. cit.

Depletion Policy and the World Energy Balance

Table 33: *OPEC members' 'proven' crude oil reserves at end 1978 and reserves to production ratio*

	Crude oil reserves million barrels	Reserves to production ratio (a)
Algeria	6,300	15
Ecuador	1,170	16
Gabon	1,970	26
Indonesia	10,200	17
Iran	59,000	31
Iraq (b)	32,100	34
Kuwait (c)	69,440	89
Libya	24,300	34
Nigeria	18,200	26
Qatar	4,000	23
Saudi Arabia (c)	168,940	56
UAE	31,316	47
Venezuela	18,000	23
Total OPEC	444,936	41

(a) Ratio of reserves at end 1978 to total production during 1978.
(b) Some estimates give Iraq's reserves much higher.
(c) Includes 50 per cent of Neutral Zone's reserves.
SOURCE: OPEC, *Annual Report 1978*, Appendix 1, p. iv.

against a cumulative production of about 60 billion barrels.'[1] Whereas Middle East oil reserves could maintain 135 years of production in 1957, this had fallen to 38 in 1977.[2] The 1970s thus saw an annual addition to reserves within OPEC, before production, of 4 billion barrels. Even if this rate of addition were maintained until 2000, curent OPEC production would result in a reduction in the ratio of reserves to production from over forty to under thirty. Even assuming ultimately recoverable reserves of 550 billion barrels, by the year 2000 OPEC members in general – though there would be differences between individual countries – would be facing a decline in production, with total exhaustion of reserves in the second

---

1 Fadhil Al-Chalabi, 'The Concept of Conservation in OPEC Member Countries', *OPEC Review*, Autumn 1979, p. 18.
2 Ibid.

quarter of the twenty-first century.¹ Studies of individual countries make it seem unlikely that OPEC will by 2000 have created an industrial base capable of providing an income to replace oil; this is the core of the OPEC case for a stricter regulation of the rate of exhaustion.

This conclusion is further supported by an analysis of the reserves to production ratios of individual countries in Table 33. They are derived from an OPEC publication and differ somewhat, though not substantially, from those given earlier in the text. It can be seen that there are only three countries with a ratio of reserves to production above that for OPEC: namely, Kuwait, Saudi Arabia and the United Arab Emirates (which in this context means Abu Dhabi). Of the other countries the highest reserves to production ratios are those of Iraq and Libya. All the other OPEC countries thus clearly face or will soon face a position where their production is constrained. This is evident from a comparison of the information on reserves in Table 33 with that on recent production in Table 34; on the basis of such a comparison estimates of the likely output for the late 1980s have been made and are contained in Table 34.

Ecuador and Gabon are by a substantial margin the smallest producers within OPEC, and have the lowest levels of reserves. They also have two of the lowest ratios of reserves to production. It is most unlikely that they will be able to maintain an increased level of oil production, and they will certainly face a fall in output before 2000. The production level for the late 1980s has, optimistically, been put at 0.2 mbd for each of them. Qatar similarly has a low level of reserves which are insufficient long to maintain the current level of output. Chapter 8 showed that Qatari production has already been reduced from its peak of 0.6 mbd in 1973. It is clear that unless new reserves are quickly found, production will fall below the existing level of 0.5 mbd.

At the end of 1978 Algeria had the lowest reserves to production ratio in the whole of OPEC. Production in 1979

1 See, Ali M. Jaidah, 'OPEC and the Future Oil Supply', *OPEC Review*, Volume I, Number 8, December 1977.

Table 34: *OPEC crude oil production*

	Million barrels per day		
	1978	1979	Projected output late 1980's (a)
Algeria	1.2	1.1	0.5 to 1.0
Ecuador	0.2	0.2	0.2
Gabon	0.2	0.2	0.2
Indonesia	1.6	1.6	less than 1.6
Iran	5.2	3.1	1.0 to 2.5
Iraq	2.6	3.4	3.5 to 4.0
Kuwait (b)	2.1	2.5	1.0 to 1.5
Libya	2.0	2.1	1.7 to 2.2
Nigeria	1.9	2.3	2.0 to 2.5
Qatar	0.5	0.5	less than 0.5
Saudi Arabia (b)	8.3	9.5	8.0 to 11.0
UAE	1.8	1.8	1.6 to 2.0
Venezuela	2.2	2.4	2.2 to 2.4
Total OPEC	29.8	30.7	24.0 to 31.6

(a) Author's estimates; see text for explanation.
(b) Includes share of Neutral Zone.
SOURCES: OPEC, *Annual Report 1978*, Appendix 1, p. vi; *Petroleum Economist*, Volume XLVII, Number 3, March 1980, p. 135; see also earlier chapters and text of Chapter 12.

was below the levels of 1977 and 1978, and she finds herself in much the same position as Ecuador, Gabon and Qatar. As stated in Chapter 9 on Algeria, in early 1980 Algeria had already announced export cutbacks of 15 per cent. SONATRACH had planned 1980 production of 1.0 to 1.1 mbd, which, with an allocation of 0.3 mbd for domestic refineries, would leave 0.7 to 0.8 mbd for exports. It was expected that the export reduction of 0.1 mbd would lead to similar reductions in production.[1] The combined effects of limited reserves and present cutbacks would permit production in the late 1980s of up to at most, 1.0 mbd, as specified in Table 34, and possibly as little as 0.5 mbd.

1 *Middle East Economic Survey*, Volume XXIII, Number 24, 31 March 1980, p. 8.

Indonesia is another OPEC member with limited reserves. At the end of 1978 her reserves to production ratio was only 17, and this had been falling as new discoveries failed to make up for current production. Consequently, production fell from 1.69 mbd in 1977 to 1.60 mbd in 1979. Indonesia's main problem is that, while there is considerable success in drilling, most of the reservoirs are small with a short productive life. Thus, although the Five Year Plan which came into operation in April 1979 optimistically forecast a build-up in production to 1.83 mbd in 1983, it seems much more likely that the next ten years will see a decline in capacity, with a maximum output by the late 1980s of 1.6 mbd.

Nigeria has substantially greater reserves than any of the countries already discussed. Indeed, provided Nigeria achieves the necessary investment in exploration and development she should be able to maintain the 1979 capacity of 2.3 mbd throughout the 1980s. The most likely constraint on the exports of Nigerian crude is a rising level of domestic demand. There may, in addition, be an imposed policy to control the rate of depletion to ensure that production can continue at present levels for the rest of this century, such a policy being acceptable because a rising oil price enables revenues to rise with unchanged production. A pointer towards this policy was provided on 1st August 1979, when the Nigerian National Petroleum Corporation cut production from 2.4 mbd to 2.15 mbd. With the uncertainties about future Nigerian production, the figure in Table 34 for the late 1980s is put at between 2.0 and 2.5 mbd. Thus, the likely production in the late 1980s for these six countries with low reserves is less than 6.0 mbd; and depending on Nigerian production, the total could be nearer 5.0 mbd than 6.0 mbd.

Venezuela is in a somewhat different category from these countries, although it has the same level of reserves as Nigeria and a low reserves to production ratio. Chapter 4 showed how Venezuelan production has fallen from its peak in the early 1970s and indeed, after having recovered to 2.4 mbd in 1979, that 2.2 mbd was the target for 1980. However, Venezuela is the only OPEC member which is placing its long-

Depletion Policy and the World Energy Balance

term hopes for oil production on the non-conventional oil of the Orinoco belt. At this stage any estimate of the ultimate reserves in the Orinoco would be speculative. But the vast sums involved in the development of the Orinoco make it most unlikely that production from the Orinoco will exceed 1 mbd by 1990. Thus for the longer term Venezuela may be able to maintain her oil production, despite the limited level of conventional reserves shown in Table 33. In the late 1980s she will be able to attain the production of 2.2 to 2.4 mbd only if success is achieved both in developing the Orinoco and maintaining conventional output.

The remaining OPEC members, Iran, Iraq, Kuwait, Libya, Saudi Arabia and the United Arab Emirates, have major reserves of crude oil, holding between them 87 per cent of OPEC's total reserves. Thus the key to future OPEC production depends on the depletion policies which these countries will pursue. The firmest prediction can probably be made in relation to Iran. Chapter 5 showed how production under the Shah was between 5.0 and 6.0 mbd during the 1970s, Iran's capacity probably amounting to 7.0 mbd. By the autumn of 1979 production under the new regime of Ayatollah Khomeini had fallen to 3.5 mbd. Since then production has fallen further, to around 1.7 mbd by mid-1980. The main cause of the reduction was probably the natural deterioration of the reservoirs and general production facilities. This deterioration could continue. Accordingly, the estimate given for Iran for the late 1980s is between 1.0 mbd and 2.5 mbd.

Iraq is one of the few members of OPEC which is adopting a policy of expanding its oil production. It was seen in Chapter 9 how Iraq was quick to take advantage of the fall in Iranian production in 1979, and increase its own production. Although the production of 3.4 mbd in 1979 was below capacity of 4.0 mbd, it was argued that this was not the result of a deliberate depletion policy. Iraq seems set to raise production further, and she will be aided in that by the fact that the reserve estimates in Table 33 may be conservative. Nevertheless, she does not have a reserves to production ratio of the size of Kuwait, and the rising price of oil helps increase revenues, so that production above 4.0 mbd by the late 1980s seems unlikely.

Kuwait, Libya, Saudi Arabia and the UAE not only have large reserves but also have substantial surplus revenues. Thus they are in a position to reduce oil production without curtailing their expenditure on development, domestic consumption and foreign aid. Libya has maintained production around 2.0 mbd since 1976, and was not among those countries which increased production in 1979 to counter the fall in Iranian exports. She is able to finance her domestic development expenditure and gross fixed capital formation with less than 100 per cent of current oil revenues and so is able to pursue a policy of regulating depletion. Chapter 11 showed how since the revolution of September 1969 she has taken a much more aggressive stance on production cutbacks and price increases. The Libyan Minister of Oil announced a reduction in supply from 1 April 1980, from 2.1 to 1.75 mbd. Present indications are that Libya will pursue a stricter depletion policy than previously, and she will certainly be able to afford this if the real price of oil continues to rise. The range for Libyan production in the later 1980s is estimated at 1.7 to 2.2 mbd.

The UAE produce below their current capacity, as was seen in Chapter 8, yet Table 33 shows that they have a reserves to production ratio of nearly fifty. The limit which Abu Dhabi, having nearly all the oil, has placed on production is a reflection of the problem of investing, internally or externally, any additional revenue. Although the presence of foreign companies, owning 40 per cent of the equity, might be thought to make for quick extraction, there appears in fact to have been a policy of regulating the rate of depletion. Indeed, according to one report Abu Dhabi had already cut back by 110,000 bd in 1980, but does not intend to go below the current 1.375 mbd.[1] As will be seen with Saudi Arabia below, the UAE may be inclined to keep production higher

---

1 *Middle East Economic Survey*, Volume XXIII, Number 23, 24 March 1980, p.2.

than it would otherwise be because of support from, and a wish not to harm economically, the industrialised countries, especially the United States. Thus the projected output for the late 1980s is put at 1.6 mbd to as high as 2.0 mbd, although the demands for revenue alone would permit lower production.

Kuwait is the OPEC country which is now taking the most important and determined stand on oil depletion policy. It became clear that the production of 2.5 mbd in 1979, increased from 2.1 mbd in 1978 as part of a coordinated effort by several OPEC members to offset the fall in Iranian output, was far more than Kuwait needed for her internal expenditures or to finance foreign aid programmes. As a result Kuwait has been accumulating foreign assets, the real value of which has been eroded by inflation, as shown in Chapter 11. That a reduction of production is now a major part of official policy is shown by the decision taken in the autumn of 1979 to curtail output to 1.5 mbd. As part of the policy, and also in a move to sell more oil directly themselves, the Kuwaiti Government has reduced the allocations to the three multi-nationals, BP, Gulf and Shell, from 1.31 mbd to 0.3 mbd. The future output in Table 34 has been put at between 1.0 and 1.5 mbd, a further increase in output being unlikely.

The future production of Saudi Arabia is seen by many, particularly the authors of the conventional studies mentioned above, as the key to future OPEC production. But given the restrictions on output which this book has argued may be imposed by other countries, any plausible range of Saudi production will still leave OPEC producing less than 30 mbd in the late 1980s. The constraints on Saudi production have been discussed in Chapter 6. Saudi revenue requirements to finance development and industrialisation, current consumption, purchases of arms and foreign aid necessitate a production of 6.0 to 7.0 mbd. However, if production were much greater than about 7.0 mbd, as at present, Saudi Arabia would, at least temporarily, be acquiring foreign assets which, as in the case of Kuwait, have tended to lose their value in real terms. In the short run, then, the Saudis would be better off if they were to forego this extra production, and leave the oil in the ground. There is no doubt that Saudi Arabia produces above this level because she believes it is in her best interests to continue to supply the industrialised countries with sufficient oil, at a price which does not rise rapidly in real terms. Continued production of 8.0 to 10.0 mbd, however, depends not only on political stability within Saudi Arabia, but also on western support for Saudi political objectives such as a Palestinian nation, and at least a degree of success in industrialising at such a rapid pace. Without the political benefits, Saudi Arabia may feel that development could be better achieved at a slower pace, thus requiring a lower level of oil production. It seems unlikely that Saudi production in the late 1980s will lie much outside the range of 8.0 to 11.0 mbd — the higher figure being improbable.

The effect of these differing national policies among OPEC members on oil depletion is to produce a total OPEC oil output for the late 1980s of between 24 and 32 mbd. Present indications are that production will be nearer 24 mbd than 32 mbd, and if outside the predicted range, is more likely to be under it than above it. It was the argument of Chapter 11 that OPEC is becoming increasingly aware of the importance of conserving oil in the ground, and with this awareness a greater cohesiveness may develop. The Kuwaiti Minister for Oil has argued that 'a shift in the production policies of oil producers . . . to production policies based on a long range vision for the economic and social transformation of their societies' is inevitable. Further, this policy 'implicitly entails, however, in the long run a reduction in the oil production by OPEC, almost to half its existing level'.[1] Such a large reduction, below 20.0 mbd, may not be achieved quickly. And there are, it has been seen, international pressures to sustain production; sometimes, though not always, domestic political pressures run in the same direction.

How does the projection of OPEC production of 24 to 32 mbd compare with other forecasts? Table 35 shows a number

---

1 H. E. Shaikh Ali Khalifa Al-Sabah, op. cit.

Table 35: *Comparison of alternative OPEC crude oil supply forecasts for 1985*

	Million barrels per day					
	Author's estimate	EIA range(a)	CIA (b)	CRS (c)	WAES (d)	OECD (e)
OPEC production	24–32	35–49	49	43	38	39

SOURCES: (a) Energy Information Administration, United States Department of Energy, 1977.
(b) *The International Energy Situation: Outlook to 1985,* April 1977, median forecast.
(c) Congressional Research Service, *Project Interdependence,* June 1977.
(d) Workshop on Alternative Energy Strategies, *Global Energy Prospects 1985-2000,* May 1977 average of cases presented.
(e) OECD, *World Energy Outlook,* April 1977, reference case.

of other forecasts of OPEC production for 1985, as well as the projection for the later 1980s based on this book and shown in Table 34. It can be seen that all the other projections are several million barrels a day above the top of the range from Table 34. The basic cause of this difference is that a study of broad magnitudes cannot take into account changes in attitude in different countries. Nor can an approach which estimates world energy demand and non-OPEC energy supplies, and regards the difference between the two as what OPEC 'should' produce at any time, have any correspondence with what OPEC is likely to produce. Further, from the viewpoint of the world energy balance one of the most important elements is OPEC exports. These depend not only on production in OPEC countries but also on their domestic consumption, which, it has been seen, is increasing rapidly, a fact understated, if not overlooked, in other studies. Table 36 shows that in the decade 1968 to 1978 OPEC domestic consumption of refined products rose by over 200 per cent, to nearly 2 mbd. Earlier chapters have shown how this consumption is expected to continue to rise rapidly, partly as

Table 36: OPEC consumption of refined products 1968–1978

	Thousand barrels per day			Percentage increase
	1968	1973	1978	1968 to 1978
Algeria	27.5	60.4	104.0	278
Ecuador	19.6	29.3	56.3	187
Gabon (a)	14.8	20.2	17.0	15
Indonesia	102.7	162.4	303.1	195
Iran	149.2	259.0	500.0	235
Iraq	64.9	85.2	175.8	171
Kuwait	11.8	16.9	36.8	212
Libya	18.5	32.7	69.8	277
Nigeria	23.3	48.9	154.1	561
Qatar	1.3	2.7	7.6	485
Saudi Arabia	27.6	68.2	259.9	842
UAE (b)	1.4	4.6	10.9	679
Venezuela	138.0	197.0	247.7	79
Total OPEC	600.6	987.5	1,943.0	224

(a) For 1968, includes consumption of Cameroon, Central African Republic, Chad, and Congo, because of their participation in the Port Gentil refinery investments.
(b) Abu Dhabi only.
SOURCE: OPEC, *Annual Report 1978,* Appendix 1, p. xii.

a result of rising living standards and, therefore, motorisation, partly of increased industrialisation with its need for energy. Higher domestic consumption will further reduce OPEC exports below the level which has previously been expected.[1]

Moving outside OPEC, Table 37 lists the production figures for the major non-OPEC producers, including the USSR and Eastern Europe, and China. It was made clear in Chapter 2 that it is only the Alaskan North Slope which has prevented production in the United States from falling more rapidly. Production from existing American fields is expected to

---

[1] See S. A. R. Kadhim and A. Al-Janabi, *Supplement of Domestic Energy Requirements in OPEC Member Countries,* OPEC Seminar, Vienna, October 1979.

Table 37: *Non-OPEC oil production*

	Million barrels per day		
	1978	1979	Estimates for late 1980s
Production:			
USA (a)	10.3	10.2	7.0–9.0
Canada (a)	1.6	1.8	1.5–1.7
UK and Norway	1.4	2.0	3.0–4.0
Mexico	1.1	1.4	3.5–4.5
Other non-communist	4.1	4.2	4.0
Net exports:			
USSR and Eastern Europe	1.7	na	−1.0 to +1.0
China	0.2	na	1.0–2.0

na/not available.
(a) Includes natural gas liquids.
SOURCES: *Petroleum Economist*, Volume XLVII, Number 3, p. 135; see also text of Chapter; *BP Statistical Review of the World Oil Industry 1978*, pp. 19, 21.

continue to fall substantially through the 1980s, and production from new fields will become increasingly important. One estimate puts US production in the late 1980s at 5 mbd from current known reserves, plus 4 mbd from new fields.[1] As has tended to be the case throughout industrialised countries, more recent official US estimates are much more optimistic about domestic production, claiming that this will rise from 9.9 mbd in 1977 to 12.4 mbd in 1995.[2] The failure of higher oil prices since 1973 to arrest the decline in US proven reserves, despite a much greater effort at exploration, and the cost and difficulty of finding reserves in new and more inhospitable areas are likely to make these official estimates unrealistic. US production for the late 1980s is thus put at roughly the present level – 7.0 to 9.0 mbd.

1 Robert Stobaugh and Daniel Yergin (Editors), *Energy Future, Report of the Energy Project at the Harvard Business School*, Random House, New York, 1979, p. 42.
2 Energy Information Administration, *Energy Supply and Demand in the Midterm: 1985, 1990 and 1995*, US Department of Energy, Washington, April 1979, p. 140.

Canadian production is similarly declining, and Chapter 3 showed how the ratio of reserves to production of conventional crude oil has fallen through the 1970s. The Albertan sands do provide a longer term means of raising production, but it is likely to be well into the 1990s before they can produce enough to offset the fall in output from conventional sources. For the later 1980s a production of 1.5 to 1.7 mbd seems more likely.

In the North Sea the United Kingdom and Norway face different constraints. It was seen that while production in the UK could rise to over 2.5 mbd in the mid-1980s, this might fall below the self-sufficiency requirement of about 2.0 mbd, and institutional arrangements combined with economic circumstances render difficult a regulation of the rate of depletion. The maximum likely production for the late 1980s is a little over 2.5 mbd, but it might be much less. Norway does not face the same limitation on reserves, but from a 1979 production of 0.4 mbd, the Storting has imposed a ceiling of 1.8 mbd for the late 1990s. For the late 1980s production will be around 1.0 mbd, which gives a range for total North Sea production of 3.0 to 4.0 mbd.

Mexico is an example of a country with newly found reserves, the extent of which, however, has been much exaggerated. While official policy on the rate of extraction is at present one of caution, the country is likely to be subject both to external and to internal pressures to raise output. Chapter 4 argued that Mexican production in 1982 would be little over 3.0 mbd. A production figure for the later 1980s of 3.5 to 4.5 mbd seems appropriate.

Production in communist oil-producing countries is not expected to change greatly in the 1980s.[1] China may increase production by a few million barrels a day, but will at the same time increase domestic consumption. Exports may, however, rise to as much as 2.0 mbd by the later 1980s.

Accepted wisdom on USSR oil production, prompted by a succession of CIA reports, which have in general underrated

---

1 For a study of USSR, see Daniel Park, *Oil and Gas in Comecon Countries*, Kogan Page, London 1979.

Depletion Policy and the World Energy Balance

Soviet capacity to overcome physical difficulties, is that it will decline and she will become a net importer of oil, the change in her net import/export balance amounting to 2 mbd. If correct, this forecast could tempt the Soviet Union into Iran to lay hands on Iranian oil. From this point of view the attempt to bolster the Communist regime in Afghanistan is irrelevant; it would be a distraction from the main aim – if aim it be – of reaching the Gulf, which is much more easily approached through the direct invasion of Iran.

However, a different scenario could be drawn. Soviet production has gradually shifted from the Caucasus to Western Siberia, where initially there was practically no infrastructure. It is this lack of initial infrastructure which has been the main reason behind the CIA's projection. However, it could be contended that this problem has been exaggerated – indeed, that it has been overcome. On this assumption Soviet production could have reached nearly 12.5 mbd in 1980, and could rise to 14 mbd by 1985 and 15 mbd by 1990. Production at this level would oblige her to reduce her exports to Western Europe, though this reduction could be offset by higher prices. It is the exports to the other Comecon countries which are likely to decline more and this could throw a weight on the external world.

The real price of oil in the next decade depends on how much of the gap between these projected oil supplies and energy demand, can be filled by non-oil supplies. These supplies are not part of the studies for this book, and consequently the estimates made by others are necessarily taken as given. They are reproduced, without comment, in Table 38.

Suffice it to say that for each source of energy all the studies agree that there is unlikely to be much expansion in production from present levels by the late 1980s. There are exceptions to this in nuclear power. Although there are great plans for expanding nuclear capacity, the achievement of these plans has been disappointing; consequently the expected nuclear production in the late 1980s shows the greatest shortfall from the projections made by earlier studies.

Table 38: *Energy production outside communist countries*

	Million barrels a day of oil equivalent	
	1978	Estimate for late 1980s
Gas	17.2	17.0–20.0
Coal	16.6	18.0–21.0
Water power	6.9	7.0– 8.0
Nuclear energy	2.8	5.0– 7.0
Oil (a)	52.2	44.5–57.8
Other	0	0.5
Total	95.7	92.0–113.8

(a) Including net exports from USSR and Eastern Europe, and China.
SOURCES: *BP Statistical Review of the World Oil Industry 1978*, p. 16; see text; Tables 34 and 37; see footnote 1, p. 190.

The demand for energy depends on the level of economic activity and the efficiency with which energy is used. The technical measure for efficient use is the 'energy coefficient', which is the amount by which energy consumption rises for a given increase in economic activity. It is defined as the percentage change in energy consumption divided by the percentage change in real gross national production. The historic value of the energy coefficient has been a little greater than 1: the average for the non-communist world from 1960 to 1972 was 1.02. This meant that energy consumption increased at roughly the same rate as economic activity. Following the oil 'crisis' of 1973/74 the energy coefficient has undoubtedly fallen, mainly due to greater efficiency in use, and for the present might be 0.85.[1]

Projections of future world economic growth are necessarily much more difficult to make. It is clear that future economic growth, until 2000, is most unlikely to achieve the levels of the 1960s and early 1970s. A plausible non-com-

---

1 See, Workshop on Alternative Energy Strategies, op. cit., p. 23; *The OECD Economic Outlook*, 25 July 1979, estimates an energy coefficient for the OECD from 1979 to 1985 of 0.8.

Depletion Policy and the World Energy Balance

munist world growth rate for the 1980s adequate for the purposes of this book would be 3.5 per cent a year. Total energy demand in 1978 outside communist countries was 93.3 mbd of oil equivalent.[1] Applying to this a growth rate of 3.5 per cent and an energy coefficient of 0.85 gives demand at the end of the 1980s at around 130 mbd of oil equivalent.[2]

The estimate of energy supply in the non-communist world in the later 1980s given in this book of 92.0 to 113.8 mbd of oil equivalent compares with the estimated demand of around 130 mbd of oil equivalent. The higher supply figure assumes the maximum production of other fuels set out in Table 38. The gap between the supply figure given here and estimated demand is such that oil will disappear from less valuable uses such as fuel oil, will be confined increasingly to uses of higher added value, such as gasoline and petro-chemicals, and will rise in price relatively to the prices of other things.

This rising real price will apply to energy in general, but particularly to oil. Supplies of other fuels cannot be quickly increased, but the supply of oil has in the past been capable of increased production. Thus in a situation where the supply of energy will be falling short of the demand for energy, the pressure on oil prices will be greater than on energy prices generally. The view that the real price of oil must be expected to continue to rise during the 1980s is thus reinforced. A higher real price for oil will also encourage OPEC members to produce less, as they will be able to meet their expenditures with a lower level of production. Thus, the higher price for oil and the resulting constraint on production are likely to act together to ensure a continuing rise in the real price of energy, and oil in particular, to 2000. This prospect would, in theory, be avoided only if there were enough new investments now in exploration, in enhanced recovery, and in alternative forms of energy. The time scale required for such investments is, however, so long as not to affect the prophecy made in this book.

[1] *BP Statistical Review of the World Oil Industry 1978*, p. 16.
[2] The exact calculation is $93.3 (1 + \frac{3.5 \times 0.85}{100})^{11} = 128.8$ mbd of oil equivalent.

CHAPTER 13

A Dialogue on Supply

Fuels in the ground form slowly, over millions of years, but they are rapidly exhausted. A usage of around 250 years, by only that fraction of the world which is industrialised, brings the end in sight. This rapidity of exhaustion applies particularly to oil, the widespread use of which is scarcely more than 30 years old. The purpose of this book has been to try to answer two fundamental questions. First, what forces or what institutions determine the rate of exhaustion or depletion? Second, what does the present rate of exhaustion imply for the industrialisation of parts of the world not yet industrialised?

Oil was first discovered in the United States, in Pennsylvania in 1859. The oil there is now exhausted and Pennsylvania has become almost a desert state. How did this exhaustion come to pass? In those days it was natural to leave the exploitation of the resource to the private individual. Now there exists a theory with a distinguished pedigree according to which, roughly speaking, the oil will be quickly depleted or extracted unless the owner finds a more profitable use for his funds elsewhere. Nowhere in the world has this theory been historically borne out. The oil of Pennsylvania, and later of other parts of the United States, was rapidly depleted because of the institutions of that country, particularly those relating to property.

According to United States law the owner of a plot of land owns everything beneath it and, for that matter, everything above it. Owners of adjacent plots of land could thus find themselves sitting on the same oilfield. And since oil, being a fluid, can migrate through the porous rock, one particular owner could suck the oil from his neighbour, storing it above

A Dialogue on Supply

ground if there was not enough demand for it and thus causing it to evaporate. This was sheer waste and dissipation.

This dissipation continued for eighty years until 1929, when most (though not all) of the oil-producing States assumed powers to control the rate of production in accordance with demand is foreseen. The exemplar of this kind of State authority is the Texas Railroad Commission, which ordained for each field its limit of production in order to sustain the price. In its attempt to keep up the price the Texas Railroad Commission was thus a model for OPEC. In spite of the attempt on the part of certain States to control the rate of production, it remains broadly true that the United States has been profligate in the extraction of its own oil, excluding, for example, the import of cheaper oil from the Middle East and keeping the domestic price below the world price, thus encouraging consumption.

After the First World War the major American companies began to move into the Middle East. The institutions relating to property were, however, different in the Middle East from what they were in the United States. In the Middle East the oil belonged to the ruler, who in turn gave the greater part to the State, which gave permission to a company to exploit the oil on payment of a royalty. In general, concessions were given to groups of companies to exploit large tracts of territory. Note that the concessions were to groups, seldom, if ever, to single companies. From the point of view of the companies operation in groups had the advantage of minimising the risk to any one member. And since roughly the same groups worked together in other concessions, losses in one concession could be offset against gains in another. Table 39, which follows, shows the distribution of the groups, and the equity proportion of each member, over the gamut of Middle Eastern concessions in January 1972.

Ideally a concession should be so exploited as to maximise its yield of oil (and gas) over time. It is difficult, however, for a consortium of companies, comprising members with differing, if also overlapping, interests, to have a common long-term view. Nor is there any indication that they

Table 39:

	Iraq	Iran	Saudi Arabia	Kuwait	Qatar	UAE
BP	23.7	40	–	50	23.75	23.75
Royal Dutch/Shell	23.7	14	–	–	23.75	23.75
Exxon	11.875	7	30	–	11.875	11.875
Mobil	11.875	7	10	–	11.875	11.875
Gulf	–	7	–	50	–	–
Texaco	–	7	30	–	–	–
Socal	–	7	30	–	–	–
Compagnie Française des Pétroles	23.75	6	–	–	23.75	23.75

SOURCE: Dr Fadhil Al-Chalabi, *Past and Present Management Patterns of the Oil Industry in the Producing Countries.*

attempted to have a long-term view. Each company appears rather to have been concerned with meeting as fully as it could the immediate demand for oil as reflected in the demand for the various derivatives from its processing activities. As for the associated gas, since the processing took place in the consuming rather than in the producing countries, it was just flared – that is, wasted.

That the companies were more concerned with the present than the future appears to be indicated by the fact that, throughout the sixties and the early seventies, the price of crude oil and net revenues per unit of production remained constant. Had regard been paid to the fact that oil was an exhaustible resource, one would have expected the price to rise through time. In sum, the companies do not appear to have had any conscious policy for regulating the rate of depletion; on the contrary, their policy seems to have been that of meeting the demand regardless of the exhaustion of the resource.

Insofar as the companies did conserve oil in the ground, it was almost by inadvertence, when they had access to reserves the exploitation of which was not needed by the prevailing demand. For example, BP had no need to develop the Iraqi concessions when it commanded plenty of crude oil in nearby Iran and Kuwait. It thus brought upon itself and its partners in the Iraqi consortium in 1961 Law 80 – that is, in effect, the seizure by the State of the unexploited part of the concessions. This proved to be one of the beginnings of nationalisation in the Middle East, and nationalisation changed the context within which the issue of depletion has to be regarded.

There were two reasons why nationalisation came about. The first was the incursion into the Middle East in the late 1950s of independent companies – that is, companies without processing activities of their own. The 'independents' showed themselves ready to offer to the owning States more generous terms than the 'majors' and to sell on the world market at a lower price. The fall in price brought with it a reduction in the revenues received from oil by the Middle Eastern Governments. The second reason for nationalisation was the failure on the part of the companies to integrate themselves into the economies of the countries in which they were operating, the tendency to keep themselves in enclaves of their own, and to make no contribution towards development or industrialisation of the host communities. In sum, they did nothing to help the countries to 'sow' the oil – that is, convert it into some other form of capital.

The evolution of nationalisation passed rapidly through various phases, and towards ends of increasing ambition. The initial goal of the governments of the oil-producing countries was the capture of a larger share of the revenues flowing from oil. To a limited extent this goal was achieved through the instrument of taxation. The further goal was the determination by the governments of the oil-producing countries of the price to be charged and the amount to be produced. The determination of the price by the governments rather than by the companies was achieved in 1971, after an abortive meeting in Tehran between companies and governments. Finally, the aim of 'sowing' the oil was met in part by the creation of a State oil company, which could work in partnership with private companies, the relative proportions increasing over time, the ultimate apportionment envisaged being 60 per cent State, 40 per cent private. This is still the case, for example, in most of the United Arab Emirates. In most cases this proportion has now gone by the board. In other instances, the State company proceeded quickly to complete nationalisation, the private companies still being used as operators, as in Saudi Arabia, or for purposes of technical assistance. To an individual State the advantage of a State company, whether it exploited the territory's oil in whole or in part, was that it had access to greater knowledge of what was possessed, was in a better position to determine the right rate of tax if private companies were still in operation, and was able to begin the training of its own personnel instead of relying entirely on expatriates.

Nationalisation has given the relevant country the nominal right to determine the rate of production. However, 'Recent studies of the behaviour of international firms suggest . . . that . . . acquisition of ownership rights, though essential, do [sic] not in themselves constitute a sufficient index of effective national control. To quote from a case study of joint ventures in India: "the general conclusion emerging from the analysis . . . is that foreign firms have managed to dilute ownership-mix in such a way as not to cause any significant diminution in foreign control in individual cases."'[1] And it has been observed in this book that the Saudi Government, in an effort to control the rate of depletion, has had to develop an intensive monitoring activity over the operating company, Aramco. Where there is no such strict monitoring, is there really an effective Government control?

The evolution is far from being complete, the need for foreign help still being substantial, though not universal. Nor can the evolution end with just the production of oil. Logically it must proceed further towards the processing of oil and indeed towards the design and the construction of the equipment required for processing. This final end lies in the far future. The fact, however, that the final goal of the

1 Zahlan, op. cit, p. 118.

evolution has yet to be attained should not blind industrialised countries to its looming possibility.

Under the private enterprise regime the relevant geographical unit was the concession; under nationalisation, on the other hand, the relevant geographical unit is the individual country; and, since the primary purpose of nationalisation is development or industrialisation the ideal depletion policy is that which best promotes over time national development. As the capital asset of oil is depleted, so its place should be taken by some other form of capital investment, so that when the oil is exhausted the country is reasonably fully developed. 'The national interest will be optimally met when the resource is depleted at a time when the highest possible level of economic development has been attained.'[1]

The importance of development is that, by providing a country with other sources of income, it can help retard the rate of depletion. If, for example, a country could establish a petro-chemical industry, the usage of oil would not be all that great and a higher value could be obtained from the derivatives than from the crude oil itself. The rate at which crude oil is extracted could, therefore, be stemmed. And the need for stemming it is urgent.

It has been said that 'in the OPEC area . . . depletion rates are the highest in the world.'[2] In the 1950s, as far as the Middle East is concerned, there was a net addition to reserves of around 100 billion barrels. In the 1960s, when post-war oil discoveries in the Middle East reached their peak, some 180 billion barrels were added to existing reserves. By contrast, between 1971 and 1978 the total net addition to reserves was only around 30 billion barrels. In other words, 'whereas, during the 1960s, much more oil was added to reserves than was depleted, the recent past shows a complete reversal . . . : much more oil depleted than oil added.'[3] If the trend of the recent past were continued and the net addition to Middle

1 Adman A. Al-Janabi, *OPEC Review*, March 1979, Vienna, p. 39.
2 Fadhil Al-Chalabi, *OPEC Review*, Vienna, Autumn 1979, p. 18
3 Ibid.

East reserves took place at the rate of only around 4 billion barrels a year, the ratio of reserves to output would fall from just under 40 in 1977 to just over 10 by the year 2000 – about the same ratio as in the United States.

The trend may, of course, be reversed; the reasons for the decline in discoveries are not fully known. The decline may be due to natural exhaustion, as in the United States. On the other hand, it may be a legacy from the era of private enterprise, the companies hesitating to explore, either because they did not wish to overload the market or because nationalisation was casting its shadow. In country after country, with the possible exception of Saudi Arabia, as the demands of the State increased so investment by the companies fell away. Out of prudence one has to assume the worst and base a judgement on the hypothesis that the present rate of depletion is leading rapidly to exhaustion.

While the primary aim of nationalisation was the development of the country, the nationalised companies initially depleted rapidly, in the belief that the extra revenues from oil could quickly bring about industrialisation. It was soon seen, however, that rapid depletion brought to light difficulties likely to retard industrialisation – inflation, an absence of skills and infrastructure, and the need for planning the industries to be established. Accordingly, around 1973 there began to be formed a new view – namely, that the rate of depletion should be retarded so as to allow the social habits of the country gradually to adapt themselves to industrialisation. Indeed, one could go further and say that industrialisation not only required a retardation of the rate of depletion, but also faciliated it, in that it promised the country a source of revenue other than oil.

The question then arises whether industrialisation can be introduced so fast as to enable the under-developed Middle Eastern oil-producing countries to stay the rate of depletion. Under-development has been described in this book as arising from conquest by a superior technology. 'Until the sixteenth century, for example, the Arabs used to produce such varied goods as damask and damascene (from Damascus), muslin (from Mosul), gauze, alcohol, arsenic, etc, much of which was exported to Europe. By the second half of the eighteenth century, France had introduced heavy import duties on cotton yarn, and machine spinning in England had become developed; both events dealt heavy blows to the export of yarn and textiles from the Ottoman Empire *for which the export of raw cotton was finally substituted*' (author's italics).[1] This history suggests that development implies a more equal spread of technological knowledge, otherwise the lot of the underdeveloped countries is just to export their natural resources. Are the countries of the Middle East in a position to absorb technological knowledge? And are the developed countries ready to transmit it?

A clue to the answer to the first question – whether the Middle Eastern countries are in a position to absorb knowledge – may be found in the history of Egypt. Muhammad Ali (1805–48) attempted an ambitious programme of industrialisation – cotton textiles, rice milling, sugar refining, edible oils, indigo dyes, glass, printing, paper, chemicals, iron foundries, naval yards, etc. All programmes were, however, totally dependent on foreign manpower, the number of expatriates in Egypt rising from 8 – 10,000 in 1838 to 90,000 by 1881. It is true that Muhammad Ali also established an educational system, but by the 1830s there were only a few technical schools on a modest scale. With Muhammad Ali's passing the entire industrial programme collapsed, the necessary arrangements not having been made to enable the native population to absorb the technology.

Are the Middle Eastern countries of modern times substantially different? By and large, it would appear not. 'There is not a general awareness . . . in the Arab world that sound development will never be attained without satisfactory development of human resources and their derivative manpower. They (Government leaders, etc.) feel that for too long national development of the Arab world has run way ahead of the process of developing resources, and that the process of creating employment opportunities as a result of

1 Zahlan, op. cit., p. 2.

development efforts has generally run ahead of the process of preparing and training new and sophisticated skills essential for filling the job opportunities thus created. They are aware also that this problem constitutes a real stumbling block in the path towards the development of the Arab world and its individual states. For a start, it may cause postponement or elimination of many projects; over the longer run, as Arab wealth approaches exhaustion, it may mean failure to obtain industrialisation.'[1]

In 1975 OAPEC conducted a survey of the extra manpower needs of the oil industry only. No subsequent survey has been undertaken, but had there been one there is no reason to believe that the conclusions would be fundamentally different. The 1975 survey concluded that over the period 1976–81 there would be required (1) 12,000 university graduates for high level posts; (2) 38,000 middle technicians for the key supervisory role; and (3) 30,000 skilled workers, a figure which it was thought could be attained. Can the needs thus precisely formulated be met? Certainly institutions exist – Baghdad has a University of Technology; Saudi Arabia has the University of Petroleum and Minerals at Dhahran; Algeria has institutes in Bumerdas and Arzew; while Baghdad is also the site of the Arab Petroleum Institute recommended by OAPEC for the 'purpose of bringing together into one system the technical and managerial organisations in the OAPEC states.'[2] The major deficiency is likely to remain at the crucial supervisory level.

And Professor Adnan Mustafa, Professor of Physics at Damascus University, reaches pessimistic conclusions; 'The low level (0.01 to 0.8 per cent of GNP) of research and development spending in the Arab countries exhibits . . . the lack of official eagerness to enhance technological development' and 'no correlation between the size of OAPEC's oil industries and the intention to build up suitable national skills

---

[1] Mahmoud Abdel-Fadil, ed. *Papers on the Economics of Oil,* Oxford Univ. Press, 1974, p. 81.
[2] Ibid, p. 86.

exists.'[1] The outsider can reach no other conclusion than that the Arab ability to absorb technology is at present low.

The basic reason for this inability to absorb is not the responsibility of the Arab world alone. The Middle East has been deluded by the fallacy lying initially behind the creation of the World Bank – namely, the idea that money alone could conjure away under-development. Money is certainly important, but it requires as an essential complement skills, skills which can establish an infrastructure – roads, railways, ports, etc – and which can plan industrial development to the measure of the infrastructure. It is significant that, in all the oil-producing countries of the Third World examined in this book, there is only one industry – oil. What little else there is serves only the local domestic market, and is not in the long run a substitute for the export of oil.

If the Arab ability to absorb technology is low, is there then a willingness on the other side – that is, on the part of industrialisted countries, including both firms and governments – to transmit technology? First, the firms. It has been seen that in all but a few cases the oil industry, both in South America and the Middle East, has been fully nationalised, indeed at an earlier date than had previously been intended.

The answer to the question whether the industrialised world is ready to transmit technology springs naturally from the nature of the competitive private enterprise system. For under such a system a firm would fear that it might undermine its own position if it were to transfer its know-how. In the short run, indeed, it might. In the long run, however, the fact of willingness to transfer could redound to its advantage, provided it could retain its ability technologically to keep ahead of others. Firms of that calibre are, however, few.

Given the slowness to absorb knowledge and the reluctance to transfer it, the multi-national companies are likely to remain in the oil-producing countries for several more decades. Few are the countries which can claim to do with a minimum of expatriates: Mexico, which has always prided

---

1 Zahlan, op. cit., pp. 437, 438.

itself on its defiance of the outside world; Iraq, though some ten years ago the story would have been different; Kuwait; and possibly Algeria, where, though SONATRACH has only a 60 per cent share in the national oil, it is claimed that only around 10 per cent of the employees are expatriates. Perhaps, however, it is not the number of expatriates that matter, but the level at which they operate, and in nearly all countries the top positions remain in foreign hands.

The foreign hold is likely to be reinforced by the absence in nearly all the countries described of an engineering industry. One may be sure that the foreign partners in the five petro-chemical projects envisaged by Saudi Arabia will insist on their own equipment — indeed there is none other — and their own foreign contractors. The design and contracting knowledge transferred to the Saudis will be little. Yet the accelerated nationalisation was but a stage in a nationalist revolution. The revolution will be consummated only when the oil-producing countries are themselves able to design, construct and operate the plants they want. True, in a world in which the division of labour was complete, the external world would provide the plants and the oil-producing countries the petro-chemicals. The real world, however, is a world of nationalisms, not of an abstract division of labour.

If private firms have contributed or are contributing little, what about the governments of the industrialised world? Nearly all the main modern technologies are now in governmental or semi-governmental hands, if only because of the vast sums of money entailed. Examples are nuclear power, telecommunications, the liquefaction or gasification of coal, the manufacture of synthetic oil, solar energy, the harnessing of the tides, etc. True, not all these technologies are as yet fully developed. Yet no government in the industrialised world has stepped forward and promised to an oil-producing country a technology which could slow down the depletion of oil.

The problem was rightly summarised by the oil-producing countries at their meeting in Algiers in March 1975. It was stated there 'that in many cases obstacles to development derive from insufficient and inappropriate transfers of technology . . .' 'The transfer of technology should not be based on a division of labour in which the developing countries would produce goods of lesser technological content. An efficient transfer of technology must enable the developing countries to overcome the considerable technological lag in their economies through the manufacture in their territories of products of a high technological content, particularly in relation to the development and transformation of their natural resources.'[1] In reality, however, it has been seen that OPEC members are neither capable of absorbing very high technology, nor is the industrialised world which possesses it prepared to transfer it.

The likelihood is that the evolution towards industrialisation never will be consummated; time will not allow it. At the beginning of this century Japan was a poor country; she has now accomplished some of the greatest break-throughs in science-based industries. True, she had a policy and a plan. But she must also have been able to change her social habits. By contrast, 'Saudi Arabia and the Gulf states are embarking on a large scheme of "industrialisation" without an adequate corresponding change in social values.'[2] It would be wrong to describe the whole of Islam as one, containing no differences. But there are common features. One has been described in these words: 'The atomism, discreteness and concreteness of . . . Arab thought . . . enabled them [sic] to develop the experimental methods. But the bold imaginative questioning of the nature of things which leads to the fundamental theories at the basis of experimental and technological advance remained foreign to them.'[3] It is the marriage of the questioning of the nature of things with revealed knowledge which is the change in 'social values' needed for accelerating the process of industrialisation. Clearly, the change cannot be quickly effected.

1 OPEC, *Solemn Declaration*, Conference of the Sovereigns and Heads of State of OPEC member Countries, Algiers, March 4–6, 1975, p.9.
2 Zahlan, op. cit., p. 187.
3 Zahlan, op. cit., p. 188.

There could, however, be one means of quickening the change. This would be the establishment of an Arab Centre for Technology, probably based on the OAPEC Centre for Engineering now being established in Baghdad. The functions of such a centre would be: to bargain on behalf of the Middle East with foreign suppliers of technology; to collect and distribute information; to fund projects uneconomic from the point of view of one country but possibly economic from the point of view of the whole; to underwrite risks entailed by advances in technology; and, naturally, to avoid duplication of effort between different countries.

There would be one important obstacle to the formation of such a centre – namely, the chain of Arab loyalties. An Arab's first loyalty is to his family; his second to his clan or tribe; a larger Arabism comes only third. Within a larger Arabism his first aim would be to fight for the predominance of his family or tribe, and in the fight the light of the larger Arabism could be extinguished. This essentially has been the difficulty facing OAPEC in its attempt to establish joint projects. This is why, to the Middle East, the Andean Pact, 'The most advanced subregional grouping so far established for cooperation in the transfer and development of technology,'[1] seems a far-off dream.

It is difficult not to conclude that the Middle Eastern attempt at industrialisation on any large scale will fail – small industries serving only the domestic market will no doubt survive. Assuming failure to develop industry capable of exporting on a scale comparable to oil, what are the consequences for depletion? Broadly speaking, the history of depletion falls into two distinct parts. In Part One, the discovery of oil leads to rapid extraction – witness, for example, the United Kingdom – the rapidity of extraction being the fruit of the combined greed of the producer and the consumer. In Part Two, on the other hand, when the immediate appetite of the producer has been sated, he wishes to prolong his supply; the greed of the consumer, however,

1 Zahlan, op. cit., p. 120.

persists, producer and consumer drift farther apart, and no dialogue is possible.

There can, of course, be slight variations around this broad theme. A country wishing to conserve its oil may not be able to do so, either because of the earlier rapid rate of extraction or because of a decline in investment – see, for example, Venezuela. A country foreseeing a short life for its reserves may try and make hay while the sun shines – Algeria is an example, although the presence of large gas reserves in that country complicates the issue. A country foreseeing for its reserves a longer life may be inclined to take things a little more easily – Iraq, for instance. Both Algeria and Iraq have Socialist regimes – quick or slow depletion does not, therefore, depend on the nature of the party regime in power. Yet again, a country wishing to conserve may be under pressure to deplete fast, the internal pressure to mitigate poverty and the external pressure of demand from a neighbouring country – for example, Mexico.

The question, however, is whether the rate of depletion is faster under a private company or under a government. And the answer given by history is clear, though with some qualification. Depletion is in general faster under a company than under a government which wishes to industrialise its country. In the 1950s, when control was in the hands of the major producing companies, their attitude towards depletion was one of caution in an attempt to maintain the price. Once, however, the independent producers had broken in on their domain, they became more heedless. Governments, on the other hand, have undergone the opposite transformation. Careless at first, they have subsequently become more prudent. The control of production is now nominally with governments, although where there are private companies as partners the latter will probably press for quicker extraction. The failure of industrialisation forecast in this book will induce a greater governmental emphasis on a more regulated rate of depletion. This evolution in turn will bring forward earlier than has hitherto been predicted a scarcity of oil, its elimination, therefore, from less valuable uses, and an increase in its real price, coupled with an attempt to bring to an equal level the price of gas when expressed in terms of heating equivalent. This attempt has already been seen on the part of Algeria, Mexico, Norway and Abu Dhabi.

The treatment has so far been confined to developing countries. But there are also developed countries which produce oil, and two of them – Canada and the United Kingdom – are undergoing a process of 'de-industrialisation'. Now de-industrialisation is not an easy concept to grasp; it seems unimaginable that a country, having become industrialised, should somehow fall back. Yet de-industrialisation is a historical reality. It happened in India in the last century, when cotton imports displaced the domestically made product and led to the pauperisation of the country. It is the same phenomenon which is now happening to Canada and the United Kingdom – the exports of manufactures being displaced by higher volumes of imports. Can oil save the de-industrialising countries? No, for oil does not touch the crux of the problem – their relative technological backwardness.

As far as the United Kingdom is concerned, oil has so far served, if not to conceal, then to mitigate the adverse balance of payments. Nor will it perform any other service for the country without a deliberate effort to redress the technological backwardness – of which there is no sign – and without an extension of the oil reserves to facilitate the technological redressment. This extension is not possible given that the proportion of North Sea oil reserves owned by the national company – BNOC – is not more than 25 per cent, while 60 per cent is in the hands of American companies with doubtful monitoring of their depletion policy. The likely life of the North Sea oil reserves is around 20 years; with the inevitable exhaustion of the reserves the United Kingdom, lulled meanwhile by an illusion, will return to the same intractable problem as before.

The conclusion for the developed oil-producing countries is thus the same as for the developing oil producers. Oil will not provide the surplus to enable the developing countries to industrialise; nor will it help the developed oil-producing countries to re-industrialise. By extension oil will not transform the Third World countries which are not producers of oil; the financial help received from OPEC countries is lavish, but the help is based on the same premiss as the World Bank - namely, that money is all. It is in fact only a partial help to the solution of an enormous problem.

It would be wrong to end a book on a pessimistic, if to the author a realistic, note. One has a duty to suggest something positive. And there is only one possible positive suggestion that one can make. There has been much talk of a dialogue between developing oil-producing countries and developed countries. Such talk way back in 1973–74 proved abortive, partly because a dialogue savoured of a confrontation and because the world was obsessed with the wrong problem – price.

An agenda for a dialogue has been laid down by the Brandt Commission.[1] The agenda contains items which would be entirely acceptable by oil-producing countries, others which would not. Nobody would object, for example, to the recommendation that 'All major energy-consuming countries will specifically commit themselves to agreed targets to hold down their consumption of oil and other energy.' On the other hand, a recommendation to the effect that 'An eventual agreement could include price indexation related to world inflation' begs the question that the rate of inflation relevant to oil-producing countries, being based in the main on capital goods, may be different from some average of the rate of increase in 'world inflation'. Similarly, the recommendation that 'oil-exporting countries, developing and industrialised, will assure levels of production' does not accord with the policy of the oil-producing countries of the Third World to adapt their extraction to their likely pace of 'development'.

It follows that a dialogue, and it can take place only between OPEC and OECD, must be concerned with mutual help and that the subject should be not the resultant, price,

1 *North-South: a Programme for Survival,* Report of the Independent Commission on International Development Issues, Pan Books, London, 1980, p. 279.

but the underlying cause, supply. Help by the developed countries means not only curtailment of demand, but also the transfer of technology particularly to facilitate the evolution of new energy forms. For example, the United States is now developing the technology for the large-scale production of solar energy; but the technology requires much capital. Why not then give it to Saudi Arabia which has plenty of capital and sun? If it is not in the nature of private firms to transfer technology, why do not governments take on the task?

As for the developing oil-producing countries, what is required from them is a statement of the rate at which they can deplete consistently with the changes which can be made in social habits, changes which cannot be hurried. Such a dialogue should at least produce a meeting of minds even if it results in no concrete policies. The case for a meeting on supply has been reinforced by the OPEC meeting in Algiers in the spring of 1980, in that it discussed the question of the control of production, a subject which national pride had previously avoided. If OPEC were indeed to resort at this stage to the idea put forward by Venezuela, when the Organisation was founded, the problem of supply would be aggravated and the position of the industrialised countries worsened.

The case for a dialogue was forcibly put by the Brandt Commission: 'The industrialised and the oil-exporting countries should reach agreement on their respective additional roles . . . it is essential that these two groups of countries join forces to transform this potential crisis into a new opportunity for co-operation – in the common interest.'[1] This recommendation was considered by the leaders of the industrialised countries at their meeting at Venice in June 1980 – and turned down. 'President Carter and Mrs Thatcher are rumoured to have squashed the enthusiasm of President Giscard d'Estaing at Venice for talks with OPEC because of fears what OPEC would demand in exchange.'[2]

The industrialised countries have consistently seen the oil-producing countries merely as the exporters of a natural

1 *North-South*, op. cit., pp. 279–280.
2 *The Guardian*, London, June 27, 1980, p. 15.

resource, just as in the last century India was seen merely as an exporter of raw cotton. They have shown themselves ignorant of the relationship between the supply of oil and the desire for development. To refuse a dialogue is to add to the ignorance an unwillingness even to try to understand.

Oil has given part of the developing world an opportunity to develop. Unless this is understood and appropriate help is given in the form of technology, the opportunity will soon have been missed and gone for ever. The oil-producing countries will become increasingly Socialist (Iran is an ominous example) and President Carter and Mrs Thatcher will have played their part in increasing Socialist influence throughout the Third World.

Bibliography

Books on oil are innumerable. The books and articles included in this bibliography are those which have been of most use to the author in the writing of the book, and which are likely to be of most value to the reader interested in exploring the subject further. They are also probably the most important.

GENERAL: DEVELOPMENT AND DEPLETION

Abdel-Fadil, Mahmoud, (editor), *Papers on the Economics of Oil*, Oxford University Press, 1974.

Aburdene, Odeh, *Middle East Economic Survey*, Volume XXII, No. 6, 27 November 1978, "The impact of inflation and currency fluctuations on OPEC's dollar assets during the period 1974–1978".

Al-Chalabi, Fadhil, *OPEC Review*, Autumn 1979, "The Concept of Conservation in OPEC Member Countries".

Al-Chalabi and Al-Janabi, *The Journal of Energy and Development*, Spring 1979, "Optimum Production and Pricing Policies".

Al-Janabi, Adran, *OPEC Review*, Volume III, No. 1, March 1979, "Production and Depletion Policies in OPEC".

Al-Janabi, Adran, *Middle East Economic Survey*, Vol. XXIII, No. 5, 19 November 1979, "Equilibrium of external balances between oil-producing countries and industrialised countries".

Al-Otaiba, Mana Saeed, *OPEC and the Petroleum Industry*, Croom Helm, London, 1976.

Baumol, W. J., *American Economic Review*, September 1968, Vol. 58, "On the social rate of discount".

Blackaby, Frank, *De-industrialisation*, Heinemann, London, 1978.

Brandt Commission, *North-South: a programme of survival*, Report of the Independent Commission on International Development Issues, Pan Books, London, 1980.

Corradi, Alberto Quiros, *Foreign Affairs*, Summer 1979, "Energy and the Exercise of Power".

Energy Information Administration, *Annual Report to Congress*, Volume II, 1977, United States Department of Energy, Washington, 1978.

Fisher, A. C. and Krutilla, J. V., *Quarterly Journal of Economics*, 1975, Volume LXXXIX, "Resource Conservation, Environmental Preservation, and the Rate of Discount".

Furtada, Gelso, *Economic Development of Latin America*, Cambridge University Press, 1978.

Griffin, Keith, *Underdevelopment in Spanish America*, Allen and Unwin, London, 1969.

Hotelling, H., *Journal of Political Economy*, April 1931, No. 39. "The Economics of Exhaustible Resources".

Huntingdon, S. P., *Political Order in Changing Societies*, Yale University Press, 1968.

Jaidah, Ali M., *OPEC Review*, Volume I, No. 8, December 1977, "OPEC and the Future Oil Supply".

Kadhim S. A. R. and Al-Janabi, A., *Supplement for Domestic Energy Requirement in OPEC Member Countires*, OPEC Seminar, Vienna, October 1979.

Kemp, Tom, *Historical Patterns of Industrialisation*, Longman, London, 1978.

Lewis, Bernard, *Islam in History*, Alcove, London, 1973.

Madujibeya, S. A., *African Affairs*, Volume 75, No. 300, July 1976, "Oil and Nigeria's Economic Development".

Middle East Economic Survey, Volume XXII, Number 48, 17 September 1979, "Conceptual Perspective for a Long-Range Oil Policy".

Nashashibi, Hikmat Sh., *Euromoney*, August 1974, "Other Ways to Recycle Oil Surpluses".

Nordhaus, W. D., *Brookings Papers on Economic Activity*, 1973, No. 3, "The Allocation of Energy Resources".

OAPEC, *Secretary General's Annual Report*, 1977 and 1978, Kuwait.

OECD, *Energy Prospects to 1985*, Volumes I and II, Paris, 1974.

OECD, *World Energy Outlook*, Paris, 1977.

OPEC, *The Statute of the Organisation of the Petroleum Exporting Countries*, Vienna, February 1978.

Park, Daniel, *Oil and Gas in Comecon Countries*, Kogan Page, London, 1979.

Pearce, D. W., and Rose J. (editors), *The Economics of Natural Resource Depletion*, Macmillan, 1975.

Ray, G. F., *National Institute Economic Review*, No. 82, November 1977, "The 'Real' Price of Crude Oil".

Robinson, Joan, *Aspects of Development and Underdevelopment*, Cambridge University Press, 1979.

Sayigh, Yusif A., *The Economies of the Arab World. Development since 1945*, Croom Helm, 1978

Shihata, I. F., and Mabro R., *World Development*, Volume 7, "The OPEC Aid Record".

Solow, R. M., *American Economic Review*, May 1974, Volume LXIV, No. 2, "The Economics of Resources or the Resources of Economics".

Stobaugh, Robert and Yergin, Daniel (editors), *Energy Future, Report of the Energy Project at the Harvard Business School*, Random House, New York, 1979.

Stork, Joe, *Middle East Oil and the Energy Crisis*, Monthly Review Press, 1975.

Workshop on Alternative Energy Strategies, Report of, *Energy: Global Prospects 1985–2000*, McGraw-Hill, New York, 1977.

World Conference, *World Energy: looking ahead to 2020*, IPC, 1977.

GENERAL: STATISTICAL PUBLICATIONS

OAPEC, *Annual Statistical Report*, Kuwait.

OPEC, *Annual Statistical Bulletin*.

OPEC, *Annual Report*.

United Nations, *World Energy Supplies 1950–1974*, Statistical Papers, Series I, No. 19, New York, 1976.

United Nations, *World Energy Supplies, 1973–1978*, Statistical papers, Series I, No. 22, New York, 1979.

Index

Page references in italic type indicates tables

Abu Dhabi (*see also* United Arab Emirates), 121–2, 172, 196, 200, *204*
Abu Safa field, 119
ADNOC, 122
Afghanistan, Soviet interest in, 207
agriculture, as source of surplus, 15, 89
Ahwaz pipe mill, 75
Akins, James, 84–5
Alberta, 40–8 *passim*
Algeria, 133, 142–50, 220
   development and industrialisation, 146–9
   domestic consumption of refined products, *204*
   and France, 142–5, 146–7, 149
   gas reserves, 145–6
   oil reserves, 145, *195*
   oil revenues, *17*
   production and exports, 30, *147*, 196–7, *197*
   reserves to production ratio, *195*, 196–7
   shared fields, 29
   technical institutes, 218
Ali, Muhammad, 217
Al-Otaibi, Dr Mana Saeed, 128
Al-Sabah, Shaikh Ali Khalifa, 193–4
aluminium, 125, 126, 127
AMPTC (Arab Maritime Petroleum Company), 130
Andean Pact, 59, 64–6
Anglo-Iranian company, 70–1
APICORP (Arab Petroleum Investments Corporation), 130
APSC (Arab Petroleum Services Company), 130
Arab Maritime Petroleum Company (AMPTC), 130
Arab Petroleum Institute, 218
Arab Petroleum Investments Corporation (APICORP), 130
Arab Petroleum Services Company (APSC), 130
Arab Shipbuilding and Repair Yard Company (ASRY), 127, 130, 131
Arabian American Oil Company (Aramco), 29, 87–9, 119
Arabian Light Crude oil, 19, 173, 175, *179, 180*
Arabian Oil Company, 86–7
Aramco (Arabian American Oil Company), 29, 87–9, 119
ASRY (Arab Shipbuilding and Repair Yard Company), 127, 130, 131

Athabasca oil sands, 43
Atomic Energy Authority, 164, 165

Baghdad University of Technology, 218
Bahrain, 118
  development and industrialisation, 124, 125, 126–8, 131
  and oil companies, 51, 121
  oil reserves, 118–20, 131
  production, 118–20
  and Saudi Arabia, 29, 119, 120, 127–8
Bank of England, 162
Bethlehem Steel, 58
BNOC (British National Oil Corporation), 154–5, 160–4, 224
Bolivia, 59n.
Boumédienne, President, 142
BP (British Petroleum), 18, 71, 102, 122, *212*, 212–13
Brandt Commission, 193, 225, 226
Brazil, 11, 50
Britain, *see* United Kingdom
British National Oil Corporation (BNOC), 154–5, 160–4, 224
British Petroleum (BP), 18, 71, 102, 122, *212*, 212–13

Canada, 38–49, *205*, 206, 224
Canadian National Energy Board, 41
Cardenas, Lazaro, 14–15
Carter, President, 226, 227
CEPE, 64
CFP (Compagnie Française des Pétroles), 144, 145, *212*

China, 204, *205*, 206
CIA, 206–7
Colombia, 59n.
Compagnie Française des Pétroles (CFP), 144, 145, *212*
companies, *see* oil companies
conservation, *see under* oil
copper, in Zambia, 12–13
corporation tax, 155–6
CVG, 58

de-industrialisation, 47, 165–7, 224
depletion rate, *see* production *under* oil
development, industrialisation, 13–25, 47–8, 58, 89, 167, 190, 215–22, 227; *see also* technology; *also* development *under individual countries*
DMA, 122
Dubai (*see also* United Arab Emirates), 121, 125, 127

Ecuador, 50, 59n., 63–6
  development and industrialisation, 64–6
  domestic consumption of refined products, *204*
  and oil companies, 63–4
  oil reserves, *195*, 196
  oil revenues, *17*
  OPEC joined by, 63, 172
  production, *65, 197*
  reserves to production ratio, *195*, 196
Egypt, 30, 217
ELF–ERAP, 72, 144, 145
energy coefficient, 208
enhanced recovery, 36–7

ENI, 72
ERAP, *see* ELF–ERAP
Estaing, President Giscard d', 226
exchange rate, changes in, and North Sea oil revenues, 158
expertise, *see* technology
Exxon, 87, *212*

FONADE, 64
Forties field, 152
France, and Algeria, 142–5, 146–7, 149

Gabon
   domestic consumption of refined products, *204*
   oil reserves, *195*, 196
   oil revenues, *17*
   OPEC joined by, 172
   production, *197*
   reserves to production ratio, *195*, 196
gas, 124–5
   in Algeria, 145–6
   in Canada, 44
   in Gulf States, 124–6
Getty Company, 86, 103
gold, 11
Great Britain, *see* United Kingdom
Guayana, 58–9
Gulf, 102, *212*

Hardwicke, Robert E., 33
Holland, 166–7
Hot Oil Act, 33
Hotelling, H., 21
Hussein, Saddam, 140

IEA (International Energy Agency), 20, 124

independent oil companies, 32, 172, 213
Indonesia, 198
   domestic consumption of refined products, *204*
   oil prices, *176*
   oil reserves, *195*, 198
   oil revenues, *17*
   OPEC joined by, 172
   production, *197*, 198
   reserves to production ratio, 195, 198
industrialisation, *see* development
interest rate, and depletion rate, 21–2
International Energy Agency (IEA), 20, 124
'invisible hand', theory of, 27
IPC (Iraq Petroleum Company), 134–6
Iran, 24, 67–82
   development and industrialisation, 67–70, 72–8, 97
   domestic consumption of refined products, *204*
   foreign assets, 185
   founder member of OPEC, 172
   Khomeini revolution, 72, 78–82,
   and oil companies, 51, 70–2
   oil prices, *176*
   oil reserves, *195*, 199
   oil revenues, *17*
   Pahlavi dynasty, 67–78 *passim*
   production and exports, *73, 74*, 80, *197*, 199
   reserves to production ratio, *195*

Iraq, 133, 134–41, 149–50
  development and industrialisation, 138–41
  domestic consumption of refined products, *204*
  founder member of OPEC, 172
  and oil companies, 51, 134–6, 220
  oil prices, *176*
  oil reserves, 136–7, *195*, 199
  oil revenues, *17*
  production and exports, 30, 136–8, *197*, 199–200
  reserves to production ratio, 137, *195*, 196, 199–200
Iraq Petroleum Company (IPC), 134–6
Islam, 79, 81, 98–9, 133, 221

Japan, 15, 221
Japan Petroleum Trading Company, 87
Jubail, 93

Khomeini, Ayatollah, 72, 78–82
Kirkuk field, 134
Kissinger, Henry, 20, 85, 188
Kuwait, 101–16
  development and industrialisation, 106–16, 127
  domestic consumption of refined products, 204
  foreign assets, 111–14, 185
  founder member of OPEC, 172
  and oil companies, 51, 102–3, 220
  oil prices, *176*
  oil reserves, *195*, 199
  oil revenues, *17*
  production and exports, 30, 102–6, *197*, 200, 201
  reserves to production ratio, *195*, 196
  shared field, 28
Kuwait Oil Company, 102, 103

Latin America, 50–66; *see also* Brazil; Ecuador; Mexico; Venezuela
law of capture, 26
Lewis, Arthur, 57
Libya
  domestic consumption of refined products, *204*
  and oil companies, 173
  oil prices, 174, *176*
  oil reserves, *195*, 199
  oil revenues, *17*
  OPEC joined by, 172
  production, 30, 173, *197*, 200
  reserves to production ratio, *195*, 196

market forces, 22–3, 27
maximum efficient rate (MER), 30
Mexico, 50, 61–3, 219–20
  development and industrialisation, 14–15, 61–2
  oil reserves, 62, 206
  production and exports, 62–3, *63*, *205*, 206
Mobil, 87, 93, *212*
Mossadegh, Dr, 70–1, 103
Muhammad Reza Shah, *see* Pahlavi dynasty
Mustafa, Professor Adnan, 218–19

National Iranian Oil company (NIOC), 67
National Oil Account, 162–3

Index

nationalisation, 16–18, 213–16; see also oil companies, national
Nigeria, 18–19, 198
  development and industrialisation, 19
  domestic consumption of refined products, *204*
  and oil companies, 18
  oil prices, *176*
  oil reserves, *195*, 198
  oil revenues, *17*
  OPEC joined by, 18, 172
  production, *197*, 198
  reserves to production ratio, *195*
NIOC (National Iranian Oil Company), 67
Nixon, President, 34
North Sea oil, see Norway; United Kingdom
Norway, 161, 166–7, *205*, 206
nuclear power, 207

OAPEC, see Organisation of Arab Petroleum Exporting Countries
OECD, 189, 225–6
oil
  conservation, waste, 26–37, 192–6, 212–13
  first discovered, 13, 27, 210
  glut, occasional, 191
  production of, depletion rate, 20–5, 27, 186–7, 190–207, 215–16, 222–4; see also under individual countries
  recovery techniques, 36
  reserves, see under individual countries
  revenues from, special funds for, 168
  shortage, feared, 190, 223–4
  'sowing' of, 57–60, 213; see also development
oil companies, 16–18, 122–4, 211–14, 219–20
  independent, 32, 172, 213
  major, 'seven sisters', 18, 71, 172
  national, need for, 47, 161, 164–5, 170, 213–14; supervision of, 104, 187
oil field, unified control of each, 13, 26–7, 28, 37
OPEC, see Organisation of Petroleum Exporting Countries
Organisation of Arab Petroleum Exporting Countries (OAPEC), 19, 124, 129–31, 218
  and OPEC investment, 183, 185
Organisation of Petroleum Exporting Countries (OPEC), 172–89
  birth of, 172
  cohesion of, 124, 179–80, 187–8
  and development, 19–20, 179, 189, 226–7
  dialogue with West, need for, 226–7
  High Court of, 29
  members of, 172
  objectives of, 172–3
  and oil production, conservation, depletion rate, 102, 186–7, 192–204, 215

OPEC (cont.)
  oil revenues and investments of members, 17, 32, 179–86
  and prices, 19, 32, 32–3, 72, 173–9
  Third World, aid for, 187–8, 225
  see also individual members
Orinoco oil belt, 53–4, 199

Pahlavi dynasty (Reza Shah and Muhammad Reza Shah), 67–78 *passim*
Pemex, 62
Peru, 59n.
Petro-Canada, 46–7
petro-chemical plants, 93–5, 128, 129
petroleum revenue tax (PRT), 156, 159
Petroleum and Submarine Pipelines Act 1975, 160, 163
Petromin, 87, 89
primary recovery, 36
prorationing, 30–2
PRT (petroleum revenue tax), 156, 159
Public Accounts Committee of House of Commons, 155–6

Qatar, 118
  development and industrialisation, 125–6, 127, 131
  domestic consumption of refined products, 204
  and oil companies, 121, 123–4
  oil reserves, 120, 131, *195*
  oil revenues, *17* 128
  OPEC joined by, 172
  production and exports, 120, *121, 197*
  reserves to production ratio, *195*, 196

Ravard, General Alfonzo, 58
reserves to production ratio, 193–6; *see also under individual countries*
reservoir management, 22
Reza Shah, *see* Pahlavi dynasty
Rotterdam, 175–7
Royal Dutch/Shell, *212*
royalties, on North Sea oil, 156
Russia, *see* Soviet Union

Sahara desert, 142, 143
Saudi Arabia, 83–99
  and Bahrain, 29, 119, 120, 127–8
  development and industrialisation, 89–99, 127, 220
  domestic consumption of refined products, *204*
  foreign assets, 184–5
  founder member of OPEC, 172
  and oil companies, 51, 86–9
  and oil prices, 72, *176*, 179
  oil reserves, 83, 194, *195*, 199
  oil revenues, *17*
  production and exports, 84–6, *197*, 200, 201–2
  reserves to production ratio, *195*, 196
  shared fields, 28, 29, 119
secondary recovery, 36
'seven sisters', 18, 71, 172
Shi-ism, 81
Sitra refinery, 120

Socal, *see* Standard Oil
socialism, 149–50, 227
Societé Nationale de Transports et de Commercialisation de Hydrocarbures (SONATRACH), 143, 144, 145, 146, 197, 220
South America, 50–66; *see also* Brazil; Ecuador; Venezuela
Soviet Union, 15, 204, *205,* 206–7
'sowing of the oil', 57–60, 213; *see also* development
'spot' market, 175–7
Standard Oil of California (Socal), 87, 120, *212*
steel, 125, 128–9
stripper wells, 31–2
Sunnis, 81
'surplus', as basis of industrialisation, 15–16, 89
Syncrude project, 43, 48
Syria, 174

technology, expertise, transmission of, to developing countries, 122–4, 167, 189, 217–22, 226–7
Tehran Agreement, 174, 177
tertiary recovery, 36
Texaco, 64, 87, 120, *212*
Texas Railroad Commission, 31, 211
Thatcher, Margaret, 226, 227
Total-Algérie, 144
Trans-Arabian Pipeline, 173–4
Tunisia, 29
Turkish Petroleum Company, 134
Turner Valley field, 39

UAE, *see* United Arab Emirates
Umm Said, 127
United Arab Emirates (UAE), 118
  development and industrialisation, 125, 127, 131
  domestic consumption of refined products, *204*
  foreign assets, 185
  and oil companies, 121–2
  oil reserves, 120, 128, 131, *195,* 199
  oil revenues, *17,* 128, 200
  production and exports, 120, *123, 197,* 200, 200–1
  reserves to production ratio, 120, *195,* 196
United Kingdom
  iron and steel, 12
  oil, 29, 151–70, *205,* 206, 224
  OPEC investment in, *182,* 183–4, 186
United States of America, 13, 26–37, 192–3, 204–5, *205,* 210–11
  and Canada, 38, 40–1, 42, 44
  and Mexico, 62
  OPEC investment in, *182,* 183–4, 186
University of Petroleum, 218
urea, 125–6
US Steel, 58
USSR, *see* Soviet Union

Venezuela, 50–61
  development and industrialisation, 55–61
  domestic consumption, 54–5, *56, 204*
  founder member of OPEC, 172

VENEZUELA *(cont.)*
  and oil companies, 50–3
  oil prices, *176*
  oil reserves, 53–4, *195*, 198–9
  oil revenues, *17, 53*
  production and exports, 33, 53–4, *197*, 198–9
  reserves to production ratio, *195*, 198

Wafra field, 28
waste, *see* conservation *under* oil
World Bank, 183, 189, 219

Yamani, Sheikh, 86
Yanbu, 93–4

Zambia, and copper, 12–13